高等学校土木工程专业"十三五"规划教材
全国高校土木工程专业应用型本科规划推荐教材

Professional English for Civil Engineering

土木工程专业英语

夏冬桃　肖本林　主　编
曾　磊　周军文　副主编

中国建筑工业出版社
China Architecture & Building Press

图书在版编目(CIP)数据

土木工程专业英语 / 夏冬桃, 肖本林主编. —北京: 中国建筑工业出版社, 2019.3

高等学校土木工程专业"十三五"规划教材 全国高校土木工程专业应用型本科规划推荐教材

ISBN 978-7-112-23282-6

Ⅰ.①土… Ⅱ.①夏…②肖… Ⅲ.①土木工程-英语-高等学校-教材 Ⅳ.①TU

中国版本图书馆 CIP 数据核字(2019)第 025458 号

本书编者基于"兴趣是最好的老师"的理念,尝试以非传统土木工程专业英语教科书的方式,精心选择教学内容,包括:做你自己的工程师、拉伸实验、工程典范——悉尼歌剧院、9·11 事件调查——世贸大厦为什么倒塌、美国凯瑞饭店人行天桥倒塌、新加坡新世界酒店倒塌、伦敦国王十字火车站地铁大火、穿越时空——地震、绿色建筑、国际交流与合作、为成为土木工程师做准备等 12 个专题。通过丰富的教学环节,使学生深刻理解土木工程师在公众健康、公共安全、社会文化、法律法规、生态环境以及可持续发展等方面应承担的责任和义务。

本教材以任务驱动教学法为编写指导思想,采用新颖的编写方式和趣味性学习内容,每个单元可按照引导活动、听力活动、阅读活动、口语活动和写作活动开展教学,学生通过观看视频、阅读课文、词语词组记忆、课堂讨论以及课后写作,加强专业英语"听、说、读、写"综合能力的提高。教师可根据课时情况作适当调整,有选择地使用本教材。

本教材可作为土木工程专业高校本科生以及研究生专业英语课程教材,也可供从事土木工程设计、施工、管理以及研究的专业人员学习专业英语参考使用。

本书配备教学课件,有需要的读者可通过发送邮件(邮件主题请注明《土木工程专业英语》)至 jiangongkejian@163.com 索取。

责任编辑:赵 莉 王 跃
责任校对:李美娜

高等学校土木工程专业"十三五"规划教材
全国高校土木工程专业应用型本科规划推荐教材
Professional English for Civil Engineering
土木工程专业英语
夏冬桃 肖本林 主 编
曾 磊 周军文 副主编

*

中国建筑工业出版社出版、发行(北京海淀三里河路 9 号)
各地新华书店、建筑书店经销
北京红光制版公司制版
天津安泰印刷有限公司印刷

*

开本:787×1092 毫米 1/16 印张:14 字数:345 千字
2019 年 3 月第一版 2019 年 3 月第一次印刷
定价:**52.00** 元(赠课件)
ISBN 978-7-112-23282-6
(33575)

版权所有 翻印必究
如有印装质量问题,可寄本社退换
(邮政编码 100037)

Preface 前言

土木工程是建筑材料、结构设计方法以及施工工艺三者有机结合的综合性工程。细节决定成败,因为无论是微小的裂缝还是瞬间的火星都可能引发土木工程结构的重大灾难。工程师既要确保所参与工程的安全性、经济性和可操作程度,还要深刻理解土木工程师在公众健康、公共安全、社会文化、法律法规、生态环境以及可持续发展等方面应承担的责任和义务。

"授人以鱼,不如授人以渔"。编著本教材的目的旨在逐步激发学生学习专业英语的兴趣,并教给他们学习专业英语的有效方法,基于"兴趣是最好的老师"的理念,精心选择和设计教学单元,不断调动学生的思维能力,提高学生的学习热情,举一反三、循序渐进,逐步掌握土木工程一般词汇和常用术语、结构计算分析和设计的一般表达方式等,提高英文科技文献的阅读能力和写作能力,以满足日益增长的国际交流与合作的需求。

本部教材具有三个特色:

1. 知识性与趣味性兼容并举。

通过引入:做你自己的工程师,材料力学最基本的实验——拉伸实验,工程典范——悉尼歌剧院,世贸大厦为什么倒塌,重返危机现场(Seconds From Disaster),包括国王十字火车站地铁大火、凯瑞饭店人行天桥倒塌、新世界酒店倒塌与地下救援、穿越时空——地震、绿色建筑和 BIM 技术简介等作为教学主要内容,信息量大,知识丰富,生动有趣。专业词汇列表和附录内容将大量的词汇放在一起,联系记忆,便于读者学一知三。在掌握土木工程一般词汇和常用术语、土木工程结构计算分析和设计的一般表达方式的同时,扩大了知识面。

2. 积极调动读者的融合思维能力,充分发挥语言学习的联想作用。

教材内容涵盖了建筑与结构设计的奇思妙想——传承与创新、荷载与结构设计方法——细节决定成败(设计缺陷与设计变更等引发灾难)、土木工程施工与管理、灾难生还者访谈(自救、感恩与缅怀)、专家灾难调查(科学试验和数值模拟)、绿色建筑与 BIM 技术等方面的交叉融合,为学生提供大量丰富而生动的工程案例,从而体验、实践和感悟问题的情境。将专业知识融入公共英语学习的语境:调查、分析、阅读、思考与创新紧密联系,有助于读者深刻理解土木工程师在公众健康、公共安全、社会文化、法律法规、生态环境以及可持续发展等方面应承担的责任和义务。

3. 专业知识、语言学习与文化修养融于一体。

On Being Your Own Engineer 源于土力学前辈 Ralph B. Peck 先生在 The University of Illinois 优秀毕业生典礼上的演讲;Sydney Opera House, Why the Towers Fell 和 Seconds From Disaster(Unit 4 ~ Unit 8)分别来源于美国 National Geographic, 美国 PBS NOVA 和英国 Darlow Smithson Productions 制作的节目,课文由原视频记录整理而成,并分成 n 小节便于

读者学习；Leap through Time-Earthquake 源于英国 Nicholas Harris 等的同名著作；Green Buildings 源于美国自然资源保护委员会 NRDC 的 Robert Watson 的大会报告；International Cooperation and Exchange 源于 Manchester University 的官方网站和 Hong Kong Polytechnic University 的院刊等。Preparation for Being a Civil Engineer 选自沈祖炎院士主编的 INTRODUCTION OF CIVIL ENGINEERING 等。学习视频与阅读课文的同时，将专业英语与西方文化结合起来，在丰富专业语言知识的同时，扩大文化交流和生态保护的视野。

本教材以任务驱动教学法为编写指导思想，任务驱动教学法契合学生主动学习心理，更利于读者专业英语"听、说、读、写"综合能力的提高。教材中每一个单元都可以通过4个阶段来展开教学：引导活动→听力活动→口语活动→写作活动。引导活动有简单问答、口语练习、词汇理解等，逐步引导学生进入到这一单元的主题上来。听力活动包括：播放视频、阶段启发问答、学生讨论答案、重复播放视频等方式。口语活动要求学生积极参与专业词汇的理解记忆和问题讨论及小组展示活动（Presentation），教师做必要的引入和指导。写作活动由教师布置课外任务（作业），学生围绕任务撰写科技短文。

编者向中国建筑工业出版社、关心和支持本书编写的领导老师以及参考文献的作者和研究生们表示诚挚的谢意！

限于编者的水平，书中的不足之处在所难免，还需要在今后的教学和研究的工作实践中不断加以改进和完善，敬请专家和读者多多批评指正。

编者
2019 年 1 月 8 日

CONTENTS 目录

Unit 1　Introduction 绪论 ⋯⋯⋯⋯⋯⋯⋯⋯⋯⋯⋯⋯⋯⋯⋯⋯⋯⋯⋯⋯⋯⋯⋯ 1
　1.1　Introduction of Professional English for Civil Engineering
　　　 关于土木工程专业英语 ⋯⋯⋯⋯⋯⋯⋯⋯⋯⋯⋯⋯⋯⋯⋯⋯⋯⋯⋯⋯⋯ 2
　1.2　Characteristics of Scientific and Technological Writing Style
　　　 科技文体的写作特点 ⋯⋯⋯⋯⋯⋯⋯⋯⋯⋯⋯⋯⋯⋯⋯⋯⋯⋯⋯⋯⋯⋯ 3
　1.3　Network and Professional English Learning 网络与专业英语的学习 ⋯⋯⋯ 5
　1.4　Task-driven Teaching Method
　　　 任务驱动教学法 ⋯⋯⋯⋯⋯⋯⋯⋯⋯⋯⋯⋯⋯⋯⋯⋯⋯⋯⋯⋯⋯⋯⋯⋯ 7
　Where There is a Will, There is a Way 有志者事竟成 ⋯⋯⋯⋯⋯⋯⋯⋯⋯⋯ 9

Unit 2　On Being Your Own Engineer 做你自己的工程师 ⋯⋯⋯ 11
　Teaching Guidance ⋯⋯⋯⋯⋯⋯⋯⋯⋯⋯⋯⋯⋯⋯⋯⋯⋯⋯⋯⋯⋯⋯⋯⋯⋯ 12
　2.1　You Can Shape Your Own Career 你可以规划自己的职业 ⋯⋯⋯⋯⋯⋯⋯ 12
　2.2　Opportunities Favor the Prepared Mind 机会总是青睐有准备的人 ⋯⋯⋯⋯ 12
　2.3　Civil Engineering Projects Exist Out in the Field & Society
　　　 土木工程项目在生产现场 ⋯⋯⋯⋯⋯⋯⋯⋯⋯⋯⋯⋯⋯⋯⋯⋯⋯⋯⋯⋯ 13
　2.4　Details Often Make or Break a Project 细节决定成败 ⋯⋯⋯⋯⋯⋯⋯⋯⋯ 13
　2.5　You Ought to Avoid Being a Job-hopper 应该避免频繁跳槽 ⋯⋯⋯⋯⋯⋯ 14
　2.6　How Can You Get the Varied Experience 怎样获得不同的实践经验 ⋯⋯⋯ 14
　2.7　Reasonable Balance among Your Goals in Life
　　　 合理平衡人生的多重目标 ⋯⋯⋯⋯⋯⋯⋯⋯⋯⋯⋯⋯⋯⋯⋯⋯⋯⋯⋯⋯ 14
　2.8　True Conservationists and True Ecologists 真正的环境和生态保护者 ⋯⋯ 15
　Words and Expressions ⋯⋯⋯⋯⋯⋯⋯⋯⋯⋯⋯⋯⋯⋯⋯⋯⋯⋯⋯⋯⋯⋯⋯ 15
　Translation Examples ⋯⋯⋯⋯⋯⋯⋯⋯⋯⋯⋯⋯⋯⋯⋯⋯⋯⋯⋯⋯⋯⋯⋯⋯ 17
　Activities—Discussion & Speaking ⋯⋯⋯⋯⋯⋯⋯⋯⋯⋯⋯⋯⋯⋯⋯⋯⋯⋯ 18
　Further Reading and Activities：Learn From Famous Scientists
　延伸阅读：向著名科学家学习 ⋯⋯⋯⋯⋯⋯⋯⋯⋯⋯⋯⋯⋯⋯⋯⋯⋯⋯⋯⋯ 18

Unit 3　The Tensile Test 拉伸实验 ⋯⋯⋯⋯⋯⋯⋯⋯⋯⋯⋯⋯⋯⋯ 23
　Teaching Guidance ⋯⋯⋯⋯⋯⋯⋯⋯⋯⋯⋯⋯⋯⋯⋯⋯⋯⋯⋯⋯⋯⋯⋯⋯⋯ 24
　3.1　Introduction to Mechanics of Materials 材料力学简介 ⋯⋯⋯⋯⋯⋯⋯⋯ 24
　3.2　The Task of a Tensile Test 拉伸实验的任务 ⋯⋯⋯⋯⋯⋯⋯⋯⋯⋯⋯⋯ 24
　3.3　The Typical Shape of the Stress-strain Diagram 典型的应力-应变图示 ⋯⋯ 25
　3.4　Necking of a Bar in Tension 拉杆颈缩 ⋯⋯⋯⋯⋯⋯⋯⋯⋯⋯⋯⋯⋯⋯⋯ 25

3.5 The Typical Stress-strain Curve for Structural Steel
典型的钢材应力-应变曲线图 ………………………………………… 26
3.6 The Typical Stress-strain Diagram of the Most Common Structural
Metal in Use 其他常用结构材料的应力-应变图示 …………………… 26
Words and Expressions ……………………………………………………… 27
Notes ………………………………………………………………………… 28
Translation Examples ……………………………………………………… 29
Reading Comprehension …………………………………………………… 29
Activities—Discussion, Speaking & Writing ……………………………… 31
Further Reading: Mechanical Properties of Ductile and Brittle Materials
延伸阅读：延性材料和脆性材料力学性能的比较 ………………………… 32

Unit 4　Sydney Opera House 悉尼歌剧院 …………………… 35
Teaching Guidance for Watching, Listening & Reading …………………… 36
4.1 A Landmark Building in Sydney 悉尼的地标建筑 …………………… 36
4.2 Inspiration Ⅰ: A Collapsible Toy & the Technique of Post-tensioning
第一个灵感：一种可折叠的玩具与后张拉技术 ……………………… 37
4.3 Inspiration Ⅱ: A Peel of Fruit (Orange) & the Magical Space
第二个灵感：橘子瓣与魔法空间 ……………………………………… 39
4.4 Inspiration Ⅲ: A Glue for False Teeth & the Precast Segments Concrete
第三个灵感：假牙的粘胶与预制混凝土构件 ………………………… 40
4.5 Inspiration Ⅳ: Gas Mask in First World War & the Large Glass Windows
第四个灵感：第一次世界大战的防毒面具与大玻璃窗户 …………… 42
4.6 Inspiration Ⅴ: Egyptian Pharaoh's Chest & the Complex Shape
Inside the Concert Hall
第五个灵感：古埃及的法老柜子与音乐大厅内部的复杂形态 ……… 44
4.7 Inspiration Ⅵ: A Copper Bottom Sailing Ship & the Air-conditioning System
第六个灵感：一艘铜底帆船与空调系统 ……………………………… 45
Words and Expressions ……………………………………………………… 49
Translation Examples ……………………………………………………… 51
Activities—Discussion, Speaking & Writing ……………………………… 52

Unit 5　Why The Towers Fell? 世贸大厦为什么倒塌？ ……… 55
Teaching Guidance for Watching, Listening & Reading …………………… 56
5.1 The Quest for 9·11 Attacks 社会各界对9·11事件的探究 ………… 56
5.2 The Structural and Fire-resistant Design of WTC (World Trade Center)
世贸大厦的结构布置和防火设计 ……………………………………… 57
5.3 The First Unthinkable Tragedy in 1993　1993年第一次恐怖袭击 …… 59
5.4 The Test Came on September 11, 2001
2001年9月11日，恐怖袭击爆发 ……………………………………… 60

5.5　Was the Fuel Load Considered in the Design?
　　　设计中是否曾考虑燃油荷载？ ·· 61
5.6　Whatever You Have, You Have to Try! 无论你有什么，你都必须尝试 ··· 61
5.7　Firefighters' Rescue Without Fear of Danger 消防队员英勇无畏的救援 ······ 63
5.8　What's Called a "Progressive Collapse"? 什么是"连续倒塌"？ ········ 63
5.9　What was Specific Steel Components' Failure?
　　　钢部件的破坏模式是什么？ ··· 67
5.10　What Does This Disaster Tell Us about the Safety of all Tall Buildings?
　　　这场灾难留下哪些关于高层建筑安全问题的警示？ ····················· 68
Words and Expressions ·· 69
Translation Examples ·· 70
Activities—Discussion, Speaking & Writing ·· 71

Unit 6　Hotel Skywalk Collapse 饭店人行天桥倒塌 ············· 75
Teaching Guidance for Watching, Listening & Reading ································ 76
6.1　It is One of the City's Most Spectacular Buildings
　　　最引人注目的建筑之一 ·· 76
6.2　The Tea Dance Competition 茶舞会 ··· 77
6.3　The Walkways Collapsed 人行天桥坍塌 ······································ 78
6.4　Over 2 Hours after the Collapse 灾后 2 小时 ································ 79
6.5　Seven and Half Hours after the Collapse 灾后 7 个半小时 ············ 80
6.6　Almost 10 Hours after the Collapse 灾后 10 小时 ························ 80
6.7　Is it Due to the Spread of the Construction or the Faulty Materials?
　　　倒塌原因是偷工减料还是建材不合格？ ····································· 81
6.8　Is the Load too Large? Is it Due to the Harmony Vibration?
　　　是负载过重？还是和谐振动？ ··· 82
6.9　Why was the Box Beam Connection Point Failed?
　　　为什么箱梁连接点失效了？ ·· 84
6.10　The Experimental Test Showed That Connections were the Cause of the
　　　Disaster 试验揭示了连接点导致了灾难的发生 ····························· 85
6.11　Whose Fault is the Fatal Flaw of the Skywalk?
　　　谁该为天桥的致命缺陷负主要责任？ ··· 86
6.12　ASCE Rewrote Its Rules
　　　美国土木工程师协会重写了规章制度 ··· 87
Words and Expressions ·· 88
Translation Examples ·· 89
Activities—Discussion, Speaking & Writing ·· 90

Unit 7　Hotel Collapse Singapore 新加坡酒店倒塌 ············· 97
Teaching Guidance for Watching, Listening & Reading ································ 98

7.1　Staff in Hotel New World 新世界酒店大楼的工作人员 ………… 98
7.2　There were Problems in the Basement Car Park
　　　地下停车库突发事故 ……………………………………………… 99
7.3　An Unstoppable Collapse was Settled 酒店大楼轰然倒塌 ……… 100
7.4　Emergency Workers Started Arriving 救援队员迅速抵达 ……… 101
7.5　About an Hour after the Collapse 灾后约一小时 ……………… 101
7.6　12 Hours after the Disaster 灾后 12 小时 ………………………… 101
7.7　Almost 2 Days Have They Been Trapped in the Rubble
　　　他们被困在废墟中近 2 天 ………………………………………… 103
7.8　The Cause of the Collapse May be an Explosion
　　　坍塌的原因可能是爆炸 …………………………………………… 104
7.9　The Cause of the Collapse May be the Instability of the Foundation
　　　坍塌的原因可能是不稳定的地基 ………………………………… 105
7.10　The Cause of the Collapse May be the Impact of the Construction of
　　　 the Subway 倒塌的原因可能是受到地铁施工的影响 …………… 106
7.11　The Cause of the Collapse May be the Micro Crack of the Pillar
　　　 坍塌的原因可能是支柱的微裂缝所致 …………………………… 107
7.12　Analysis of the Causes of Micro Cracks 微裂缝产生的原因分析 … 107
7.13　Conclusions of Investigators 调查组的结论 ……………………… 109
　　Words and Expressions ………………………………………………… 110
　　Translation Examples …………………………………………………… 111
　　Activities—Discussion, Speaking & Writing ………………………… 112

Unit 8　King's Cross Fire 国王十字火车站地铁大火 …………… 117

Teaching Guidance for Watching, Listening & Reading ……………… 118
8.1　Rush Hour on the City Center 市中心的高峰时刻 ……………… 118
8.2　A Small Flame on One of the Wooden Steps 木台阶上惊现小火星 … 119
8.3　Red Watch Arrived 消防队迅速抵达 ……………………………… 119
8.4　The Flame was about a Meter and Half High 火星飞溅至 1.5 米高 … 120
8.5　Open the Gate. Hello! 快请打开大门! ……………………………… 121
8.6　Whether it was an Arson or a Terrorist Attack? 是纵火还是恐怖袭击? … 123
8.7　Whether it was Caused by a Thrown Cigarette?
　　　是由乘客扔香烟引发的火灾? …………………………………… 124
8.8　Whether it was Caused by "Piston Effect"?
　　　是由"活塞效应"导致的火灾? ………………………………… 126
8.9　Find out the Reason of the Eruptive Fire 勘查起火的原因 ……… 127
8.10　Simulation Test Confirmed the "Trench Effect"
　　　 数值模拟证实了"沟槽效应" …………………………………… 127
8.11　Conclusions of Investigators 调查组的结论 ……………………… 129
　　Words and Expressions ………………………………………………… 129

Translation Examples ·· 131
Activities—Discussion, Speaking & Writing ·· 131

Unit 9　Leap through Time – Earthquake 穿越时空-地震 ······ 135
Teaching Guidance ·· 136
9.1　About 3000 Years Ago 大约 3000 年前 ·· 136
9.2　A Hundred Years Ago 大约 100 年前 ·· 137
9.3　A Few Years Ago 大约几年前 ··· 137
9.4　Later That Day 那天晚些时候 ··· 137
9.5　Seconds Later 几秒以后 ·· 138
9.6　At the Same Time 同时 ··· 138
9.7　A Few Minutes Later 几分钟后 ··· 139
9.8　Twenty Minutes Later 二十分钟后 ·· 139
9.9　Several Hours Later 几个小时后 ··· 140
9.10　The Next Morning 次日凌晨 ·· 140
9.11　Today, a Few Years Later 几年后的今天 ······································· 140
Words and Expressions ·· 141
Further Reading ·· 141
Activities—Discussion, Speaking & Writing ·· 144

Unit 10　Green Buildings 绿色建筑 ·· 145
Teaching Guidance ·· 146
10.1　Environmental Impact of Buildings 建筑对环境的影响 ···················· 146
10.2　What is "Green" Design? 什么是"绿色"设计? ······························· 146
10.3　Green Building Assessment Systems 绿色建筑评估体系 ·················· 147
10.4　Leadership in Energy & Environmental Design (LEED)
　　　绿色建筑评估认证标准体系 ·· 148
10.5　Non-Economic Benefits of Green Building
　　　绿色建筑的非经济性效益 ··· 150
10.6　Green Buildings in China 中国的绿色建筑 ····································· 150
10.7　Green Building Features 绿色建筑特征 ·· 150
Further Reading: Introduction to BIM (Building Information Modeling)
延伸阅读：BIM（建筑信息模型）介绍 ·· 151
Activities—Discussion, Speaking & Writing ·· 154

Unit 11　International Cooperation and Exchange
　　　　　国际交流与合作 ·· 157
Teaching Guidance ·· 158
11.1　MACE of the University of Manchester 曼彻斯特大学机械、
　　　航天与土木工程学院 ··· 158

11. 2　The 15th International Symposium on Structural Engineering
　　　第十五届国际结构工程专家研讨会 …………………………………………… 161
11. 3　Events on The Hong Kong Polytechnic University
　　　香港理工大学活动剪影 ……………………………………………………… 164

Unit 12　Preparation for Being a Civil Engineer
　　　　　为成为土木工程师做准备 …………………………………………… 167
　Teaching Guidance ………………………………………………………………… 168
　12. 1　What Kind of Knowledge is Necessary for a Civil Engineer?
　　　　土木工程师需要什么样的知识? …………………………………………… 168
　12. 2　What Can the University Education Provide for Students?
　　　　大学教育能为学生提供什么? ……………………………………………… 170
　12. 3　What Abilities Shall a Future Civil Engineer Possess?
　　　　未来土木工程师应具备什么样的能力? …………………………………… 172
　12. 4　How to Match the Demands of the Program Education?
　　　　你如何与项目教育的要求相匹配? ………………………………………… 174
　12. 5　It's Never Too Late to Learn 活到老，学到老 ……………………………… 176
　Further Reading：Project Management Responsibilities
　延伸阅读：工程管理部职责 ………………………………………………………… 176
　Activities—Discussion & Speaking ………………………………………………… 179

Appendix, References & Acknowledgements
　　　　　附录，参考文献 & 致谢 …………………………………………………… 181
　Appendix 1：General Terms
　　　　　　附录1：土木工程一般术语 ………………………………………… 182
　Appendix 2：Canonical Terms
　　　　　　附录2：土木工程规范术语 ………………………………………… 186
　Appendix 3：Punctuation Marks and Typefaces
　　　　　　附录3：标点符号和字体 …………………………………………… 204
　Appendix 4：Weights and Measures
　　　　　　附录4：度量 ………………………………………………………… 205
　Appendix 5：Numerals and Mathematical Symbols
　　　　　　附录5：数和数学符号 ……………………………………………… 208
　References 参考文献 ……………………………………………………………… 210
　Acknowledgements 致谢 ………………………………………………………… 211

Unit 1

Introduction

1.1　Introduction of Professional English for Civil Engineering
　　关于土木工程专业英语

　　2018年6月24日，土木工程专业正式纳入我国工程教育专业认证体系及《华盛顿协议》名单，此举将有助于土木工程专业技术人员跨境流动和执业，支撑"一带一路"国家战略的实施。专业认证的核心理念是成果导向教育（Outcome Based Education，OBE）。OBE强调如下4个问题：学什么？为什么学？如何学？如何评价学习效果？

一、课程性质与目标

　　土木工程专业英语对阅读土木工程专业英文原版书籍和文章感兴趣的学生所开设，使其能对土木工程学科与技术领域及其相关行业的国际状况有基本了解，并能表达自己的观点；旨在进一步提高学生阅读理解和综合分析、熟悉专业词汇、了解科技文体的能力，能够运用图纸、图表和文字对土木工程的复杂工程问题进行有效表达，进一步提高学生听、说、读、写的综合能力，以满足日益增长的国际科技交流与合作的需求。

二、教学基本内容

　　工程师既要确保所参与工程的安全性、经济性和可操作程度，又要深刻理解土木工程师在公众健康、公共安全、社会文化、法律法规、生态环境以及可持续性发展等方面应承担的责任和义务。本教材建议学时为32学时，教师可根据实际情况作适当调整，各个单元学时分配如下：

　　　Unit 1：Introduction　　　　　　　　　　　　　2学时
　　　Unit 2：On Being Your Own Engineer　　　　　2～3学时
　　　Unit 3：The Tensile Test　　　　　　　　　　　2～3学时
　　　Unit 4：Sydney Opera House　　　　　　　　　3～4学时（包含视频播放）
　　　Unit 5：Why the Towers Fell?　　　　　　　　　3～4学时（包含视频播放）
　　　Unit 6：Hotel skywalk Collapse　　　　　　　　3～4学时（包含视频播放）
　　　Unit 7：Hotel Collapse Singapore　　　　　　　3～4学时（包含视频播放）
　　　Unit 8：King's Cross Fire　　　　　　　　　　　3～4学时（包含视频播放）
　　　Unit 9：Leap through Time—Earthquake　　　　3～4学时
　　　Unit 10：Green Buildings　　　　　　　　　　　2～3学时
　　　Unit 11：International Cooperation and Exchange　2～3学时
　　　Unit 12：Preparation for Being a Civil Engineer　　2～3学时

三、课程学习方法

　　基于任务驱动教学法，学生可以在视频学习、联想（拓展）学习、课堂（小组）讨论以及分组展示等多种学习过程中，不断地提升自己"听、说、读、写"的专业英语综合能力。

联想学习就是利用联系查找和扩展词汇，举一反三。其涉及三种方法：辐射联想、推理联想和交叉联想。辐射的方法是以一个表示话题的核心词汇为中心，通过分类属性、特征和功能等，联想与其相关的内容。推理的方法是从一个已知的词出发，利用逻辑关系找到它的反义词、近义词、上义词和下义词等。交叉的方法是由一个话题领域联想到另一个话题领域。

例如：说到土木工程（civil engineering），自然想到，建筑物（building）、公路（highways）、地铁（subway）、桥梁（bridge）、机场（airport）等，以及规划（planning）、设计（design）、施工（construction）、管理（management）等。

例如：由 building（建筑物、房屋、大楼），联想出：你居住的大楼（the building you live in）、建筑物的分类（the different kinds of building）、房屋外部（内部）（outside or inside of a building）、房屋的维修和加固（maintenance and reinforcement of a building）、房屋的买卖和出租（buying, selling and renting a building）等。

四、课程的考核环节及课程目标达成度自评方式

考核成绩以平时成绩与期末试卷成绩综合构成。其中平时作业（论文或研读报告）、课堂讨论、分组展示、考勤综合为平时成绩占 50%，期末试卷考试成绩占 50%。本课程重点支撑 2 个毕业要求指标点，指标点对应的课程教学目标、达成途径和评价依据如下：

指标 1：能够运用图纸、图表和文字对土木工程的复杂工程问题进行有效表达。

指标 2：掌握和应用一门工具（英语），能对土木工程学科与技术领域及其相关行业的国际状况有基本了解，并能表达自己的观点。

教学目标：了解专业英语在口语表达中的应用；掌握土木工程一般术语和常用术语、结构计算分析以及结构设计的常用表达等；掌握土木工程专家介绍的常用文体和句型；掌握科技文献的常用术语和句型，具备一定的阅读能力和写作常识。

达成途径：以任务驱动教学法为主，结合视频教学以及课堂讲解。课堂主要通过列举法、分角色讲解以及分组讨论和分组演示，课后布置相应的任务（练习）作业训练巩固所学知识。

评价依据和方式：平时成绩和期末相关知识点得分。每次任务完成的平时作业、课堂表现、考勤的平时综合成绩占 50%，期末考试占 50%。

1.2 Characteristics of Scientific and Technological Writing Style 科技文体的写作特点

一、科技论文的基本特征

科技论文是在科学研究、科学实验的基础上，对自然科学和专业技术领域里的某些现象或问题进行专题研究，运用概念、判断、推理、证明或反驳等逻辑思维手段，通过分析和阐述，揭示这些现象和问题的本质及其规律性而撰写成的论文。科技论文应该具有科学性、首创性、逻辑性和有效性，这也就构成了科技论文的基本特征。

二、科技论文的分类

从目前期刊所刊登的科技论文来看主要涉及以下 5 类：

第一类是论证型——对基础性科学命题的论述与证明，或对提出的新的设想原理、模型、材料、工艺等进行理论分析，使其完善、补充或修正。

第二类是科技报告型——描述一项科学技术研究的结果或进展，或一项技术研究试验和评价的结果，或论述某项科学技术问题的现状和发展的文件。

第三类是发现、发明型——记述被发现事物或事件的背景、现象、本质、特性及其运动变化规律和人类使用这种发现前景的文章。阐述被发明的装备、系统、工具、材料、工艺、配方形式或方法的功效、性能、特点、原理及使用条件等的文章。

第四类是设计、计算型——计算机程序设计，某些系统、工程方案、产品的计算机辅助设计和优化设计以及某些过程的计算机模拟，某些产品或材料的设计或调制和配制等。

第五类是综述型——它要求在综合分析和评价已有资料基础上，提出在特定时期内有关专业课题的发展演变规律和趋势。

三、英文科技论文的特点

1. 大量使用名词化结构 (nominalization)

科技文体中，名词使用范围很广，名词化结构强调存在的事实，而非某一行为。例如：Archimedes first discovered the principle of displacement of water by solid body. 译文：阿基米德最先发现固体排水的原理。名词化结构，一方面简化了同位语从句，另一方面强调 "displacement" 这一事实。名词化的作用是使句子更加紧凑，从而原来要用两个分句表达的内容用一个简单句就可以了。

动词结构名词化的操作方法：① 选用适当的名词替代动词：a) 有些动词可通过加上名词化后缀变成名词。常用的名词化后缀有：-ment，-tion，-y，-sion，-ance，-al 等。b) 有些动词本身可作名词用。c) 没有相应名词形式的可以借助 -ing 构成名词。②动宾结构，名词化后用 of 引出原来的宾语。③主谓结构，名词化后的主语用 of 引出。④主谓宾结构，用 by 引出主语，用 of 引出宾语。⑤后接介词的动词在变成名词后保持原来的介词。

2. 广泛使用被动句 (the passive sentence)

科技文章侧重叙事推理，强调客观准确。科技文体所述的是客观规律，尽量避免使用第一或第二人称。不必说出主语的情况通常会使用被动句。例如：炉壁采用耐火砖可大大降低热耗。可采用以下两种表达：参考译文 1，The heat loss can be considerably reduced by the use of firebricks round. 参考译文 2，The use of firebricks round the boiler can considerably reduce the heat loss.

3. 常使用非限定动词 (the nonrestrictive verb)

非限制性动词包括：分词短语、分词独立结构、不定式、介词加名词短语。

4. 大量使用后置定语 (the postposition)

例句：The forces due to friction are called frictional forces.

参考译文：由于摩擦而产生的力叫摩擦力。

5. 定语从句（attributive clause）

例句：During construction, problems often arise which require design changes.

参考译文：在施工过程中，常会出现需要改变设计的问题。

定语从句的常用句型有以下几种：① It...that...句型：如 It is believed that.../据信；It is proved that.../已经证实、有人证实；It is reported that.../据报道。② 被动句型：Computers may be classified as analog and digital. ③ as 结构句型：Microcomputers are very small in size, as is shown in Fig. 5. ④ 分词短语句型：The resistance being very high, the current in the circuit was low. ⑤ 省略句结构句型：An object, once in motion, will keep on moving because of its inertia.

6. 长句较多（the long complex sentence）

例句：The effort that have been made to explain optical phenomena by mean of the hypothesis of a medium having the same physical character as an elastic solid body led, in the first instance, to the understanding of a concrete example of a medium which can transmit transverse vibration, and, at a later stage, to the definite conclusion that there is no luminiferous medium having the physical character assumed in the hypothesis.

分析：翻译这样的长句，分析句子成分是第一步，只有句子成分分析对了，理解与翻译才有可能正确。The effort 为本句的主语，led to 为谓语，the understanding 为宾语。

7. 复合词及缩略词多（the compound and the abbreviation）

复合词：full-enclosed 全封闭的；work-harden 加工硬化；crisscross 交叉着；on-and-off-the-road 路面越野两用的。

缩略词：maths = mathematics 数学；lab = laboratory 试验室；Ft = foot/feet 英尺，ASCE = American Society of Civil Engineering 美国土木工程师协会；CAD = computer aided design 计算机辅助设计。

8. 多种方法表达数量（the number and the quantity）

例句1：The factory turns out 100,000 cars every year.

例句2：A yard is three times longer than a foot.

1.3　Network and Professional English Learning 网络与专业英语的学习

一、专业文献的检索与阅读

1. 专业文献的分类

按文献载体划分为：（1）印刷型，包括印刷本（printed copy）、复印本（duplication copy）、预印本（preprint copy）。（2）缩微型，包括缩微胶卷（microfilm）、缩微胶片（microfiche）、缩微卡片（microcard）和缩微印刷片（microprint），要借助显微镜阅读。（3）机读型（machine-readable），这类文献存在于计算机储存介质中。Internet 信息也属于机读型。（4）直感资料，包括唱片、录音带、录像带、视频等。

按文献加工深度分为：(1) 一次文献（原始文献），如论文、报告等。(2) 二次文献，对一次文献用一定方法进行加工、归纳、简化、组织成为系统和便于查询的文摘类文献，即书目、题录、文摘等"检索工具"。(3) 三次文献，利用二次文献检索一次文献，并在广泛了解和掌握一次文献的内容基础上通过研究、分析、综合而编写出来的文献，如进展报告、专题述评等。

按出版类型分为：科技期刊、科技图书以及特种文献。特种文献包括科技会议文献（proceedings）、学位论文［（英）thesis，（美）dissertation］、科技报告、专利文献（patent）、政府出版物、标准文献等。

2. 期刊和会议文献的检索

(1)《美国工程索引》(Engineering Index, EI)。EI 采用主题法编制文摘，检索人员选准了主题词（按 EI 所给的工程主题词表 - Abstracts EI Thesaurus）就能像查词典一样检索主题词下的文献集合。从 EI 网站查询文献，可访问 http://www.ei.org。

(2)《美国科技会议录索引》(Index to Scientific and Technical Proceedings, ISTP)。由美国 Institute of Scientific Information 公司编辑出版，以题录形式报道多学科领域会议文献的检索。用搜索引擎 google，northernlight 或其他搜索工具，以 Institute of Scientific Information 或其简称 ISI 为关键词进行搜索，可得 ISI 的主页网址：http://www.isinet.com。

(3)《美国科学引文索引》(Science Citation Index, 简称 SCI)。它是由美国科学信息研究所（ISI）1961 年创办出版的引文数据库。SCI（科学引文索引）、EI（工程索引）、ISTP（科技会议录索引）是世界著名的三大科技文献检索系统，是国际公认的进行科学统计与科学评价的主要检索工具，其中以 SCI 最为重要。

(4) 科技报告。例如，美国国家技术情报服务局（National Technical Information Service, NTIS），其主页为 http://www.ntis.gov，美国国家科学院（National Academic of Science, NAS）及其下属的国家科学研究委员会（National Research Council, NRC），NAS 的主页为 http://www.nationalacademics.org，通过 NAS 可查阅土木工程类科技报告；美国国家标准与技术协会（National Institute of Standards and Technology, NIST），其主页为 http://www.nist.gov/，利用该主页可查阅土木工程类科技资料。

(5) 学位论文。例如，《国际学位论文摘要》是美国出版的专门检索学位论文的主要工具之一，由国际大学缩微公司（University Microfilms Internal，简称 UMI，现改名为 Bell & Howell Information and Learning Company）出版。其主页为 http://www.umi.com。

(6) 专利文献。例如，世界知识产权组织（World Intellectual Property Organization），是隶属于联合国的国际组织，其主页为 http://www.wipo.int/eng/main.htm；英国得温特公司（DERWENT），其主页为 http://www.derwent.com；美国《专利公报》由美国专利商标局（United States Patent & Trademark Office, USPTO）出版，其主页为 http://www.uspto.gov/。

(7) 标准文献。在土木工程领域，标准文献（Standard, Specification or Code）指的是关于工程项目规划、设计、施工、养护、维修等技术规范，以及相关产品的生产过程、技术规格、质量检验等技术文件。例如：国际标准化组织（International Organization for Standardization, ISO），其主页为 http://www.iso.ch；英国标准协会（British Standards Institute, BSI），其主页为 http://www.bsi.org.uk，是世界上历史最悠久的标准制定团体；

美国国家标准协会（American National Standards Institute，ANSI），其主页为 http://www.ansi.org/；欧洲标准委员会（European Committee for Standardization，CEN），其主页为 http://www.cenorm.be。

(8) 科技图书。科技图书的一般结构包括封面（cover）、扉页（title page）、版权页（copyright page）、目录（contents）、序言（preface, foreword）、致谢（acknowledgements）、符号注释（notation）、词汇表（glossary）、正文（text）、附录（appendix）、索引（index）、参考文献（references）等。

二、网络英语学习资源

1. 超级工程。《超级工程》是中央电视台推出的一部纪录片，展现了五个重大工程项目：《港珠澳大桥》、《上海中心大厦》、《北京地铁网络》、《海上巨型风机》和《超级LNG船》。超级工程成为展示强盛国力的符号标志，彰显出现代中国的时代风采。其网址可访问 http://jishi.cntv.cn/program/cjgc/。

2. 超大建筑狂想曲。本网站学习系列是以计算机合成画面为主轴，节目揭示是什么样的发明，让潜水艇、运输机、邮轮、水坝、隧道、钻井平台、观景摩天轮、望远镜、太空站和体育场，得以在尺寸和规模上更上一层楼。其网址可访问 http://www.jlpcn.net/vodhtml/345.html。

3. 重返危机现场。《重返危机现场》（Seconds From Disaster）是在国家地理频道播放的一个电视纪录片系列，由英国的 Darlow Smithson Productions 制作的节目。《重返危机现场》的最大特色是通过生还者的回想经过和官方资料，在相似场景或电脑成像下，以顺序形式重演事件发生的经过直至发生意外的一刻，并分析每场事故的起因和经过对事件的影响。其网址可访问 http://natgeotv.com/uk/seconds-from-disaster/about/。

1.4 Task-driven Teaching Method 任务驱动教学法

一、态度决定高度

土木工程专业正式纳入我国工程教育专业认证体系及《华盛顿协议》名单后，将有助于土木工程专业技术人员跨境流动和执业，支撑"一带一路"国家战略的实施。学生应充分认识到学习专业英语的使命感和必要性，这是学好专业英语的必要条件。**Attitude is altitude.**

本课程的主要教学方法是"任务驱动教学法"，即教师为学生提供体验、实践和感悟问题的情境，学生围绕任务展开学习，以任务的完成结果检验和总结学习过程。

学习了 Unit 2 "做你自己的工程师"（On Being Your Own Engineer）的演讲报告后，学生应认识到职业规划对自身发展的重要性，对自己的学习进行简单的规划，制定一个合理的计划（a sound plan），意识到"机会是给有准备的人"，掌握好专业外语的应用，可为学生今后的职业选择和进一步发展提供更多的可能性。

二、兴趣是最好的老师

在《NOVA》（Unit 5）《重返危机现场》（Unit 6～Unit 8）的视频学习过程中，教师在播放精彩的视频之前，应给学生提出相关的问题。学生应带着问题进行视听学习，边听边记录关键的专业词汇。在视频中，通过灾难调查工程师团队细致入微地分析灾难现场，抽丝剥茧；通过提出一种假设、分析判断否定假设、再提出另一种假设、再分析判断的方法，最后综合分析得出灾难调查的结论。

视频播放结束后，学生应积极用英语来回答相关问题。有一定的思路，但英文口语表达不流畅的学生，可用基本完整的句子回答问题；建议英语基础薄弱、口语较差的学生用关键词回答。小组讨论以及分组展示是锻炼口语以及培养团队意识的好机会，学生应能积极在团队中承担个体、团队成员以及负责人的角色，与其他组的同学进行有效沟通和交流，包括撰写报告和设计文稿、陈述发言等。**Mere curiosity adds wings to every step.**

在视听教学过程中，学生不仅可以在全英文的视听环境下锻炼听力，积累一些核心词汇，而且单词与视频内容的结合增加了专业英语学习的趣味性，有助于对教学内容的掌握。在灾难调查小组专家的分析思路引导下，学生可以学习到研究人员分析问题、解决问题的方法，并应用到研究生探究学习的过程中。

三、举一反三和熟能生巧

1. 描述学科领域的专家

在 Unit 2 的拓展阅读中，教师给学生展示三位科学家从事的领域和主要成就，将关于描述科学家的英文关键短语以及核心词汇及时标出。教师有针对性地进行教学练习，学习描述人物的词、短语以及句型。学生在课堂上即兴用英文描述自己的导师/某位任课老师，包括其求学经历、职称/职务、研究方法、科研成果等。学生在进行人物描述时尽量多使用范文的表达，通过举一反三的仿写，掌握专家介绍的常用文体和句型。**Practice makes perfect.**

2. 科技文献阅读与翻译

Unit 3 拉伸实验（The Tensile Test）是材料力学最基本的实验，这篇科技文章很具有代表性。该文章的整体结构（目的、方法、结果、结论）、数学/物理推导过程（试验步骤、数据采集、曲线描述）都值得学生发挥联想的作用，举一反三地学习。基于任务驱动教学的教材内容可以充分调动学生的思维能力，善于总结发现专业英语与公共英语的差异性，例如，strain harden 译为应变硬化，proportional limit 译为比例极限。课后的学习任务，在规定时间内完成一篇与学生目前专业研究方向相关的实验（试验）介绍，然后以书面形式上交作业。通过联想拓展学习和仿写小文章，可掌握科技文献的写作基本常识。

四、更上一层楼

提高了专业英语听力和基本写作的能力后，将会对专业英语产生更浓厚的学习兴趣。为了进一步提高，建议学生每天至少抽出 1 个小时来听《重返危机现场》全英文视频，直到将视频中的每一个专业核心词都理解透彻为止，并能用自己的语言介绍视频的主要内容，用流利的英语回答每个视频中提出的若干问题。通过对视频的进一步学习以及任务作业的完成，可逐步掌握土木工程一般词汇和常用术语、结构计算分析和设计的表达方式等，提高英文科技论文的阅读能力和写作能力。**Make further progress.**

通过工程典范（Unit 4）、土木工程灾难视频（Unit 5～Unit 8）以及绿色建筑（Unit 10）等专题的学习，学生了解建筑与结构设计的奇思妙想、结构工程师在建筑中的重要地位。细节决定成败，因为无论是微裂缝还是小火星都可能引起建筑结构的重大灾难。对于土木工程师于公众健康、公共安全、社会和文化以及法律等方面应承担的责任，学生应建立较深刻的感性认识和理解。

If a builder builds a house for a man and does not make its construction firm, and the house which he has built collapse and cause the death of the owner of the house, that the builder shall be put to death.

——*The Code of Hammurabi, Babylon, circa* 2000 B.C.

五、视野拓展

学生提高了专业英语"听、说、读、写"的能力，才可能参加国际会议或出国留学，才可能在国际会议中出色地宣读论文，同时有幸聆听很多学者的优秀报告，与参会学者自由交流，对当前国际上前沿的研究方向及方法和内容有更多的了解。这部分内容的介绍将在 Unit 11 中进行。

Where There is a Will, There is a Way
有志者事竟成

如果一个人只能有一个信念，那么就是：I CAN！"机会总是留给有准备的人"，应不断鼓励自己：要把握当下，相信"一分耕耘、一分收获"。因为：

今天是我们的日子（Today is made for us），每秒都很重要，分秒必争（Every minute counts），今日的决定就是明日的事实（Today's decisions are tomorrow's realities）。要乐观（Be optimistic），要下定决心（Be determined），想追求目标就要即刻行动，因为犹豫者必失良机（Those who hesitate are lost）。

专心是成功的关键（Concentration is the key to success），有志者事竟成（Where there's a will, there's a way），成功操之在我们手中（Be the master of our successful story）。

人人都能出人头地（Every man is a king！）。

Science is a powerful instrument. How is it used? Whether it is a blessing or a curse to mankind, depends on mankind and not on the instrument.

——*Albert Einstein*

Unit 2

On Being Your Own Engineer

Teaching Guidance

The occasion for this short talk was the *Civil Engineering* students Annual Awards Convocation at the University of Illinois on April 24, 1976. Parents, friends, and wives or husbands of the honor students had been invited to the *convocation*. I took the occasion to the wives or husbands as well as to the students who received the honors.

Ralph B. Peck (1983)

2.1 You Can Shape Your Own Career

Here at this University and in this Department that has trained so many outstanding civil engineers, you have achieved a standard of excellence that results in your recognition at this Honors Days *Ceremony*. It gives me the greatest pleasure to congratulate you on these achievements. Here in your undergraduate career, you have become leaders in the pursuit of engineering knowledge, the first essential step in becoming a civil engineer. Excellence in undergraduate studies correlates highly with a successful engineering career in later years. I sincerely hope that the satisfaction of a successful career continues to be yours and that these honors and recognitions that you so rightfully receive today will be only the first of many satisfactions that will come to you in your practice of civil engineering.

Yet a successful undergraduate career is not always or *inevitably* followed by leadership in your profession. In a changing world, in a *dynamic* profession such as civil engineering, how can you be sure today that you will be among the leaders of your profession 20 or 30 years from now? How can you even be sure to pick the *branch* of civil engineering, the particular kind of work that you will actually like the best or have the most *aptitude* for? Do you dare leave these matters to chance, do you dare let nature simply take its course? Nobody can predict the future and nobody can guarantee success in the future. But there are, nevertheless, many positive things you can do to shape your own career. I should like to think about some of these with you today.

2.2 Opportunities Favor the Prepared Mind

I believe every engineer, perhaps even while an undergraduate but certainly upon graduation, needs to form and follow his own plan for the development of his professional career. Perhaps it is an unpleasant thought, but I believe it is only realistic that nobody else is quite as interested in your career as you yourself should be. If you don't plan it yourself, it is quite possible that nobody will. On the other hand, there are too many factors, there are too many changes in a dynamic profession to permit laying out a fixed plan. The plan that you follow must be flexible and it must continually be evaluated. To be sure, every career depends *to some extent* on chance, on the breaks, good or bad. But if you have followed a sound plan, you will be ready for the good breaks when they come. Those who feel they have never had favorable opportunities usually have not been ready and have not even recognized opportunities when they came.

2.3 Civil Engineering Projects Exist Out in the Field & Society

Civil engineering projects don't exist in the classroom or in the office or in the *laboratory*. They exist out *in the field*, in society. They are the *highways*, the *transit systems*, the *landslides* to be *corrected*, the *waste disposal plants* to be constructed, the *bridges*, the *airports*; they have to be built by men and *machines*. In my view, nobody can be a *good designer*, a *good researcher*, a leader in the civil engineering profession unless he understands the methods and the problems of the *builders*. This understanding ought to be *firsthand*, and if you are going to get it, you have to plan for it. Without this experience in the field, your designs may be impractical, your research may be irrelevant, or your teaching may not prepare your students properly for their profession.

There are several ways in which you can get *construction experience*. One is by being an engineer for a builder, for a *constructor*. Or on the other hand, you might be an *inspector* for a *resident engineer* for the designer or *owner*. It doesn't matter in what *capacity* you work, and it doesn't taken a very long time to get *worthwhile experience* in the field, but sometime early in your career, you should plan to get it. Since the real projects are out there in the field, you will have to go where they are to get the construction experience, and you may have to *put up with* a little inconvenience in order to get it.

2.4 Details Often Make or Break a Project

Real problems of civil engineering design include both *concept* and *detail*. In fact, details often make or break a project. A beautifully designed *cantilever bridge* in Vancouver *Harbor collapsed* during construction because a few *stiffeners* were *omitted* on the webs of some *temporary supporting beams*. Spectacular failures such as this don't always follow from *neglected* details, but *poor design*, *poor engineering* often do.

I believe every civil engineer needs a *personal knowledge* of the details of his branch of civil engineering. If he is going to be a *geotechnical engineer*, for example, he needs to know among other things exactly how *borings* are made and *samples* taken under a variety of *circumstances*. If he is going to be a *structural engineer*, he needs to know how *steel structures* are actually *fabricated* and *erected*. He needs to know, in other words, the state of the *commercial art* that plays such a large part in his profession. He needs to know how things are *customarily* done so that he can tell whether, for example, a *commercially available* sampling tool will do the job at a modest *competitive price* or whether some unusual tool must be developed for the *particular requirements* of the job. So it seems to me that you should plan to get this sort of experience also: to spend some time on a *drilling rig* if you plan to be a geotechnical engineer; to work for a *steel fabricator* or in a *design office* if you intend to be a structural engineer.

2.5 You Ought to Avoid Being a Job-hopper

How can you get this varied experience, these various *components* of civil engineering that are so dissimilar? I think, for the most part, you have to do it by choosing your jobs carefully and changing your job if and when it seems necessary. You may be lucky in your very first job and go to work for an organization that designs, that *supervises constructions*, that makes its own *laboratory tests*, that supervises borings, and so on. If this should be true, you would be fortunate, but this is not usually the case. Even such an organization may tend to let you get stuck in one phase of their work, and you may have to persuade them from time to time to let you work in other parts of their activities. More likely you will have to change organization, possibly even to move to another part of the country or of the world. Unfortunately you can't order the jobs that you want, when you want them, and where you want them. But you can look at every opportunity to see if it fits in your plan and to judge if the time is right to make a change.

The breadth of experience so important in a civil engineer's background can't be obtained any other way than by a variety of jobs or a variety of activities within a given job. You owe it to yourself and to your career to see that you get this varied background. On the other hand, while you're getting this background, you ought to avoid being a *job-hopper*. Each of your *employers* will have an *investment* in you. At least for a while, when you start to work for him, he will not be getting his money's worth from you. You owe him a return on his investment, you owe him good work, and you owe staying with him a reasonable minimum time while you're getting that experience.

2.6 How Can You Get the Varied Experience

On my first real job, I had the good fortune to be working under Karl Terzaghi. He had a good many requirements, but one of the most important was that I should keep a notebook in which I should record not just what I had done that day, but what I had seen, what I had observed. When I went down into a *tunnel heading*, I should come back and *sketch* how the heading was being *executed* and how it was being *braced*. I soon discovered that very often, when I came back, I couldn't remember exactly what had gone on in the heading. I couldn't remember exactly how the bracing fit together. In other words, my eyes had seen what was going on, but my brain didn't really register. My powers of observation were poor. But as I continued to keep this notebook, I discovered that more and more I could remember what I had seen, and more and more my powers of observation developed, I recommend this to you as one way to make your experience more meaningful.

2.7 Reasonable Balance among Your Goals in Life

An investment of ten years or so after your degree, including perhaps graduate studies as well, *in accordance with* a carefully planned but flexible program, will go a long way toward assuring success in your engineering career. But there is another important aspect to be considered. Any worthwhile career is *demanding*. It makes demand on your time and effort, and also on your family. And there are other demands on your life besides your career. Your wife or your husband will have her or his own goals and even may also have a career in mind. The demands of others in your life and the *fulfillment* of their goals and careers will require cooperation, adjustment, give and take. Moves from one place to another will require leaving friends; will require that your children change schools. Tension and conflicts are inevitable and *compromise* and reason are necessary. You and your partner will need the best possible understanding. Many a marriage has foundered on the career ambitions of one or both partners and, *conversely*, many a career has *foundered* on unreasonable or non-understanding social or financial demands of the partner. There is seldom a perfect solution to this problem, but there are many good solutions. The important thing is to face up to the problems early and to keep working on them. The best engineers, I think have achieved a reasonable balance among their goals in life. Often they can truly say that their partner in life has also been their partner in their career.

2.8 True Conservationists and True Ecologists

Your generation has a most exciting *prospect*. Don't believe for a minute the *prophecies* that technology has *outlived* its usefulness. You will have, fortunately, much more to consider than technology. You will need to be true *conservationists*, true *ecologists* in the *positive sense*. You will need to be involved in the social *cost-benefit* assessments of civil engineering work above and beyond the dollar cost-benefits. Progress in these directions will be the challenge and the great achievement of your generation, and it is an exciting prospect. But to succeed, you must be *fully prepared*, not poorer, but better *grounded technically* than your *predecessors*. In next ten years, the choices you make and the experiences you get will be *crucial*. As Honor Students, you have taken the first necessary step with skill and distinction. All of us, your teachers, your partners, your husbands, wives, and friends wish you even greater success in the future. Indeed you must succeed, or this world will be a poorer place rather than a richer place in which to live.

Words and Expressions

convocation	会议，集会	laboratory	实验室，研究室
ceremony	典礼，仪式	highway	公路，大路
inevitably	不可避免地，必然地	landslide	滑坡，山崩
dynamic	动态的，动力的	correct	改正，修正
branch	分支，出现分歧	bridge	桥梁
aptitude	天资，自然倾向	airport	机场，航空站

machine	机械，机器	crucial	重要的，决定性的
builder	建筑者，施工人员	civil engineering	土木工程
firsthand	直接的，直接得来的	to some extent	在某种程度上
constructor	构造器，建造者	civil engineering project	土木工程项目
inspector	检查员，巡视员	in the field	在实地，在现场
owner	所有者，业主	transit system	运输系统
capacity	能力	waste disposal plant	污水处理厂
concept	理念，观念	good designer	好的设计师
detail	细节，详情	good researcher	良好的研究人员
harbor	海港	construction experience	施工经验
collapse	倒塌，瓦解	resident engineer	常驻工程师
stiffener	加劲杆（板、条、肋），加强杆	worthwhile experience	有价值的经验
omit	遗漏，省略	put up with	忍受，容忍
neglect	忽视，疏忽	real problem	真正的问题
boring	钻探，钻孔	cantilever bridge	悬臂桥
sample	样品，样本	temporary supporting beam	临时支撑梁
circumstance	环境，情况	poor design	不良设计
fabricate	制造，装配	poor engineering	不良工程
erect	使竖立，建造	personal knowledge	个人水平
customarily	通常，习惯上	geotechnical engineer	岩土工程师
job-hopper	跳槽者	structural engineer	结构工程师
employer	雇主，老板	steel structure	钢结构
investment	投资，投入	commercial art	商业艺术
sketch	画草图	commercially available	市场上可买到的
execute	实行，执行	competitive price	公开招标价格
brace	支柱，支撑	particular requirement	特殊要求
demanding	苛求的，要求高的	drilling rig	钻机，钻探装置
fulfillment	履行，满足	steel fabricator	钢材加工商，钢结构制造者
compromise	妥协	design office	设计室，设计办公室
conversely	相反地	component	组成部分，构件
founder	失败	supervise construction	监督施工
prospect	前景，前途	laboratory test	实验室试验
prophecy	预言	the breadth of experience	丰富的经历
outlive	比…活得长，比…经久	tunnel heading	隧道工作面，掌子面
conservationist	生态环保人士	in accordance with	依照，与…一致
ecologist	生态学家	positive sense	正义
cost-benefit	成本效益的	fully prepared	充分的准备
predecessor	前任，前辈	be grounded technically	打下牢固的技术基础

Translation Examples

[1] Civil engineering projects don't exist in the classroom or in the office or in the laboratory. They exist out in the field, in society. They are the highways, the transit systems, the landslides to be corrected, the waste disposal plants to be constructed, the bridges, the airports.

土木工程项目不在教室里、办公室里、或者实验室里，而是存在于社会生产现场，例如公路或干道、运输系统、滑坡治理、垃圾处理场建设、桥梁和机场。

[2] Real problems of civil engineering design include both concept and detail. In fact, details often make or break a project.

对于土木工程设计来说真正的难点是关于概念和细节方面的处理。事实上，细节经常决定一个工程的成败。

[3] A beautifully designed cantilever bridge in Vancouver Harbor collapsed during construction because a few stiffeners were omitted on the webs of some temporary supporting beams.

温哥华海港的一个很壮观的悬臂桥就是因为漏掉了一些用于临时支撑梁腹板中部上的加劲杆（加强杆）而断裂了。

[4] If he's going to be a geotechnical engineer, for example, he needs to know among other things exactly how borings are made and samples taken under a variety of circumstances. If he's going to be a structural engineer, he needs to know how steel structures are actually fabricated and erected.

举例来说，如果他想做一个岩土工程师，他需要知道钻孔是怎样钻出来的，怎样在地下进行采样的。如果他要做一个结构工程师，他需要知道钢结构是怎样进行工业化生产制作以及钢结构是如何安装的。

[5] When I went down into a tunnel heading, I should come back and sketch how the heading was being executed and how it was being braced.

当我下到一个隧道的掌子面，返回时需要对这个工程的施工状况做一个描述，并且记录它是怎么建造和支撑起来的。

[6] My powers of observation were poor. But as I continued to keep this notebook, I discovered that more and more I could remember what I had seen, and more and more my powers of observation developed, I recommend this to you as one way to make your experience more meaningful.

我的观察能力实在是有限。但是当我开始记录这些笔记的时候，慢慢地我发现我可以记下来我看到的东西了，我的观察力也提升了，我举这个例子也是为了告诉你们，这也是一种方法来丰富你的工作经验。

[7] An investment of ten years or so after your degree, including perhaps graduate studies as well, in accordance with a carefully planned but flexible program, will go a long way toward assuring success in your engineering career.

取得学位或者毕业以后将近十年的时间里，需要认真制定灵活的规划直到确保你们成功了。

[8] But there is another important aspect to be considered. Any worthwhile career is demanding. It makes demand on your time and effort, and also on your family.

但是另外一个很重要的方面你们需要考虑。任何一个值得的工作都是要求很高的。它可能需要你投入大量的时间和精力，而且你的家庭成员也要全力支持你。

Activities—Discussion & Speaking

Presentation

Group: 5 to 7 members
10 minutes per group (Each member should cover your part at least one or two minutes).
Clearly deliver your points of the following questions to audiences.
NEED practice (individually and together)!!
Gesture and eye contact.
Smile is always KEY!! Cover your nervousness!!

Questions for discussion and presentation

Preparation for being a civil engineer:
1. What kinds of knowledge are necessary for a civil engineer?
2. Are you able to a general description for the necessary knowledge that provides a future civil engineer with solid science and engineering basement?
3. What can the university education provide for students?
4. What abilities shall a future civil engineer possess?
5. How do you match the demands of the program education?
6. Why does an engineering student have to have good understanding for human science and the knowledge relating non-engineering fields?

Further Reading and Activities

Reading Material

Learn From Famous Scientists

The historical development of mechanics of materials is a fascinating blend of both theory and experiment—theory has pointed the way to useful results in some instances, and experiment has done so in others. Such famous persons as Leonardo da Vinci (1452~1519) and Galileo Galilei (1564~1642) performed experiments to determine the strength of wires, bars, and beams, although they did not develop adequate theories (by today's standards) to explain their test results.

Leonardo da Vinci said: " Mechanics is a mathematic paradise, because we acquired mathematics's fruit here."

1. Galileo—Father of Modern Science

Galileo(1564 ~ 1642) is an Italian astronomer, mechanist and philosopher. He was born in Pisa on Feb. 15, 1564 and died on Jan. 8, 1642 at the same place. He made a detailed study on the basic concepts of movement including the center of gravity, speed and acceleration and came up with the rigid mathematic formulas. Especially the concept of acceleration is the milestone in the history of mechanics. He once informally proposed law of inertia, which established the foundation for Newton to propose formally the first law and the second law. It can be said that Galileo is the pioneer of Newton in the settlement of the classical mechanics. Galileo also brought up with the law of resultant and the rule of the parabolic motion and set up the principle of relativity. He is the first scientist to make a lot of achievements by the telescope. He kept on fighting with the idealism and church philosophy and suggested that we should study the law of nature by specific experiments and thought that experiences are the source of theory.

伽利略（1564~1642），意大利天文学家、力学家、哲学家。1564年2月15日生于比萨，1642年1月8日卒于比萨。伽利略对运动基本概念，包括重心、速度、加速度等都作了详尽研究并给出了严格的数学表达式。尤其是加速度概念的提出，是力学史上的里程碑。伽利略曾非正式提出过惯性定律，这为牛顿正式提出第一、第二定律奠定了基础。在经典力学的创立上，可以说伽利略是牛顿的先驱。伽利略还提出过合力定律、抛物线运动规律，确立了伽利略相对性原理。他是用望远镜观测天体取得大量成果的第一位科学家。他一生坚持与唯心论和教会的经院哲学作斗争，主张用具体的实验来认识自然规律，认为经验是理论知识的源泉。

2. Isaac Newton—The Greatest Genius of all Time

(1)

Isaac Newton(1642 ~ 1727) is a great British physicist, mathematician and astronomer. He was born in a family of peasants in Lincoln on Dec. 25, 1642 and died of kidney stone in London on Mar. 20, 1727. Here are his contributions to mechanics: He made a further study on the basis of Galileo and other people, and concluded the three principles of objects' movement and made a firm foundation for mechanics. He is the discoverer of the gravitation law and set up the theoretical system of the classical mechanics. He also made profound contributions to the field of mathematics, optics and astronomy. *The math theory of natural philosophy* is his most important work. He concluded many important discoveries and study results in all his life in this book.

牛顿（1642~1727）是伟大的英国物理学家、数学家和天文学家。1642年12月25日生于林肯郡的一个农民家庭，1727年3月20日因肾结石症在伦敦逝世。在力学方面的贡献：牛顿在伽利略等人工作的基础上进行深入研究，总结出物体运动的三个基本定律，为力学奠定了坚实的基础,牛顿是万有引力定律的发现者。他还创立了经典力学理论体系,在数学、光学、天文学等方面均做出了开创性的贡献。《自然哲学的教学原理》是牛顿一生最重要的著作,该书总结了他一生许多重要的发现和研究成果。

(2)

Isaac Newton was born in 1642, he was a mathematician and a physicist. During the period of 1669~1701, he was a professor at Cambridge University. In 1689 he became an M. P. (Member of Parliament) for the University, and in 1699 he became Master of the Mint. Newton has been described as one of the greatest scientists ever. A lot of his original work was done immediately after his graduation at his parents' home of Lincoln shire, while the university was closed (1665~1667) during the Great Plague.

His first discovery was the law of gravitation, which inspired by the realization that an apple falling from a tree is attracted by the same force that holds the Moon in orbit. Newton's second major work in this period was the invention of the Calculus. Newton and Leibniz argued for some years as to who had the idea first. It is probable that they both invented the method independently. His third contribution was in the area of optics. From his work in this area he was inspired to invent the reflecting telescope (1668).

Although Newton became a Whig M. P., he made little impact in politics. He did, however, reform the coinage when he was Master of the Mint. He was President of the Royal Society from 1703 until his death. He published his second principle work, *Optics*, in 1704; he was knighted in 1705. Sir Isaac died in 1727. He is buried in Westminster Abbey.

艾萨克·牛顿出生于1642年,他是一个数学家和物理学家。在1669~1701年期间,他在剑桥大学担任教授。在1689年,他成为大学的国会议员,并在1699年,他成为了造币厂厂长。牛顿被认为是有史以来最伟大的科学家之一。他的许多原创作品都是毕业后在他的父母林肯郡的家里完成的,然而在大瘟疫期间(1665~1667),大学被关闭。

他的第一个发现就是万有引力定律,这一启发是由于认识到从树上落下的苹果被同样的力量所吸引,这种力量将月球保持在轨道上。牛顿在这一时期的第二大贡献是微积分的发明。牛顿和莱布尼兹多年来一直争论是谁第一个有这个想法的。他们很可能是独立发明了这种方法。他的第三个贡献是在光学领域。从他在这一领域的工作中,他受到了启发,发明了反射望远镜(1668年)。

尽管牛顿是一名辉格党人,但他对政治影响不大。然而,当他是铸币厂厂长时,他确实改制了硬币。他从1703年开始担任皇家科学协会的主席,直至去世。1704年,他出版了他的第二部作品《光学》,1705年被封为爵士。艾萨克爵士于1727年去世,逝世后被埋葬在威斯敏斯特教堂。

3. Stephen Prokofievitch Timoshenko—Outstanding Scientist and Educationalist

Stephen Prokofievitch Timoshenko (1878~1972) is a Russian dynamicist with American nationality. He was born in Ukraine on Dec. 23, 1878 and died in Germany on May. 29, 1972. He began his creative work between 1903 and 1906 and did research in the University in Germany every year supervised by famous scholars. He was a professor of colleges between 1907 and 1917. He

came to America in 1922 and engaged in the study of mechanics. In 1928, he founded "the mechanics department of ASME" and held various kinds of mechanics seminars periodically. He has many works on applied mechanics. Especially since the year of late 1920s he has written about 20 books applied such as *Mechanics of Materials*, *Advanced Mechanics of Materials* and *Mechanics of Structures* except that he has done some work in teaching and training masters.

铁木辛柯 S. P. （1878～1972）是美籍俄罗斯力学家。1878年12月23日生于乌克兰的什波托夫卡，1972年5月29日卒于联邦德国。1903～1906年他每年去德国格丁根大学在著名学者的指导下从事研究工作，并开始了他的创造性工作。1907～1917年在学院担任教授。1922年到美国从事力学研究工作。1928年他建立了"美国机械工程师学会力学部"，定期组织各种形式的力学讨论会。铁木辛柯在应用力学方面著述甚多，特别是自20世纪20年代末，除授课和培养研究生外，把主要精力用于编写书籍，他编写了《材料力学》、《高等材料力学》、《结构力学》等二十种著作。

Discussion, Speaking & Writing

Team-work

Form groups of four or five and discuss these questions below, give your opinions about each of the questions. Find out if others share your opinions. Give presentations group by group!

Questions for discussion and writing

How do you consider the importance of personal motivation, enthusiasm of innovation in higher education? Then work in groups of four or five and share with other members in your group of the famous person I admire (Name, nationality, age, occupation, description).

Write down the information about a famous person you admire who's engaged in civil engineering (This person can be a living scientist, professor and so on).

Exercises

Translate Timoshenko's academic works and curriculum below into English.

铁木辛柯是一位力学教育家，他主讲过很多重要的力学课程，还培养了许多研究生。除授课和培养研究生外，他还把很多精力用于编写书籍，编写了《材料力学》、《高等材料力学》、《结构力学》、《工程力学》、《高等动力学》、《弹性力学》、《弹性稳定性理论》、《工程中的振动问题》、《弹性系统的稳定性》、《板壳理论》和《材料力学史》等20多部。这些教材影响很大，被翻译为世界各国的多种文字出版，其中大部分有中文译本，有些书至今仍被采用。此外他还写了《俄国工程教育》和《自我回忆》两本书。

Unit 3
The Tensile Test

Teaching Guidance

Theoretical analyses and experimental results have equally important roles in mechanics of materials. Theories are used to derive formulas and equations for predicting mechanical behavior but these expressions cannot be used in practical design unless the physical properties of the materials are known. Such properties are available only after careful experiments have been carried out in the laboratory. Furthermore, not all practical problems are amenable to theoretical analysis alone, and in such cases physical testing is a necessity.

3.1 Introduction to Mechanics of Materials

Mechanics of materials is a *branch* of *applied mechanics* that deals with the *behavior* of *solid bodies* subjected to various types of loading. Other names for this field of study are *strength* of materials and mechanics of *deformable bodies*. The solid bodies include *bars* with *axial loads*, *shafts* in *torsion*, beams in *bending*, and *columns* in *compression*. *Beams* and *rods* are the main *research objects* of mechanics of materials. Most structures are made up from the beams and rods.

The principal objective of mechanics of materials is to determine the stresses, strains, and displacements in structures and their components due to the loads acting on them. An understanding of mechanical behavior is essential for the safe design of all types of structures, whether airplanes and antennas, buildings and bridges, machines and motors, or ships and spacecraft. That is why mechanics of materials is a basic subject in so many engineering fields. Most problems in mechanics of materials begin with an examination of the external and internal forces acting on a stable deformable body. First the loads acting on the body are defined, along with its support conditions, then reaction forces at supports and internal forces in its members or elements are determined using the basic laws of static equilibrium (provided that the body is statically determinate).

In mechanics of materials you study the stresses and strains inside real bodies, that is, bodies of finite dimensions that deform under loads. To determine the stresses and strains, we use the physical properties of the materials as well as numerous theoretical laws and concepts. Mechanics of materials provides additional essential information, based on the deformations of the body, to solve statically indeterminate problems (it not possible using the laws of static equilibrium alone).

The tasks of mechanics of materials: Under the *request* that the strength, *rigidity*, *stability* are satisfied, the necessary *theoretical foundation* and *calculation method* are offered for determining *reasonable shapes* and *dimensions*, choosing *proper materials* for the *components* at the most economic price. The *tensile test* is the basic test of the mechanics of materials.

3.2 The Task of a Tensile Test

The *relationship* between *stress* and *strain* in a particular material is determined by means of a tensile test①. A *specimen* of the material, usually in the form of② a round bar, is placed in③ a *testing machine* and subjected to tension④. The *force* on the bar and the *elongation* of the bar are *measured* as the load is increased. The *stress* in the bar is found by dividing the force by⑤ the *cross-sectional* area, and the strain is found by dividing the elongation by the length along which the elongation occurs. In this manner⑥ a complete *stress-strain diagram* can be obtained for the material.

3.3　The Typical Shape of the Stress-strain Diagram

The *typical shape* of the stress-strain diagram for structural steel is shown in Fig. 3 – 1(a) where the axial strains are *plotted* on the *horizontal axis* and the *corresponding stresses* are given by the *ordinates* to the curve *OABCDE*. From *O* to *A* the stress and strain are directly *proportional* to one another and the diagram is *linear*. Beyond point *A* the linear relationship between stress and strain no longer exists; hence the stress at *A* is called the *proportional limit*. For *low-carbon (structural) steels*, this limit is usually between 30,000 psi, and 36,000 psi, but for high-strength steels it may be much greater. With an increase in loading, the strain increases more rapidly than the stress until at point *B* a considerable *elongation* begins to occur with no *appreciable* increase in the tensile force. This phenomenon is known as *yielding* of the material, and the stress at point *B* is called the *yield point* or yield stress. In the region *BC* the material is said to have become *plastic*, and the bar may actually elongate plastically by an amount which is 10 or 15 times the elongation which occurs up to the proportional limit. ⑦At point *C* the material begins to *strain harden* and to offer additional resistance to increase in load. Thus, with further elongation the stress increases, and it reaches its maximum value, or *ultimate stress*, at point *D*. ⑧Beyond this point further stretching of the bar is accompanied by a reduction in the load, and *fracture* of the specimen finally occurs at point *E* on the diagram.

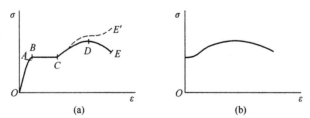

Fig. 3 – 1　Typical Stress-strain Curve for Structural Steel
(a) Pictorial Diagram (Not to Scale); (b) Diagram to Scale

3.4　Necking of a Bar in Tension

During elongation of the bar, a *lateral contraction* occurs, resulting in a decrease in the cross-sectional area of the bar. This phenomenon has no effect on the stress-strain diagram up to about point *C*, but beyond that point the decrease in area will have a noticeable effect upon the *calculated value* of stress. A pronounced *necking* of the bar occurs (see Fig. 3 – 2), and if the actual cross-sectional area at the narrow part of the neck is used in calculating σ, it will be found that the true stress-strain curve follows the dashed line *CE'* in Fig. 3 – 1(a).

Fig. 3 – 2 Necking of a Bar in Tension

Whereas the total load the bar can carry does indeed *diminish* after the ultimate stress is reached (line *DE* in Fig. 3-1a), this reduction is due to the decrease in area and not to a loss in strength of the material itself. [9] The material actually *withstands* an increase in stress up to the point of failure. For most practical purposes, however, the conventional *stress-strain curve OABC-DE* in Fig. 3-1(a), based upon the original cross-sectional area of the specimen, provides satisfactory in formation for *design purposes*.

3.5 The Typical Stress-strain Curve for Structural Steel

The diagram in Fig. 3 – 1(a) has been drawn to show the *general characteristics* of the *stress-strain curve* for steel, but its proportions are not realistic because, as already mentioned, the strain which occurs from *B* to *C* may be 15 times as great as the strain occurring from *O* to *A*. [10] Also, the strains from *C* to *E* are even greater than those from *B* to *C*. A diagram drawn in proper proportions is shown in Fig. 3 – 1(b). In this figure the strains from *O* to *A* are so small in comparison to the strains from *A* to *E* that they cannot be seen, and the linear part of the diagram appears as a *vertical* line.

3.6 The Typical Stress-strain Diagram of the Most Common Structural Metal in Use

The presence of a pronounced yield point followed by large *plastic strains* is somewhat unique to steel, which is the most common structural metal in use today. Aluminium alloys exhibit a more gradual *transition* from the linear to the *nonlinear* region, as shown by the stress-strain diagram in Fig. 3 – 3. Both steel and many aluminium alloys will undergo large strains before failure and are therefore classified as *ductile*. On the other hand materials that are *brittle* fail at relatively low values of strain (see Fig. 3 – 4). Examples include *ceramics*, *cast iron*, *concrete*, certain *metallic alloys*, and glass.

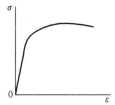

Fig. 3 – 3 Typical Stress-strain Curve for Structural Aluminium Alloy

Fig. 3 – 4 Typical Stress-strain Curve for a Brittle Material

Diagrams analogous to those in tension may also be obtained for various materials in compression, and such *characteristic stresses* as the *proportional limit*, the yield point, and the *ultimate stress* can be established.① For steel it is found that the proportional limit and the yield stress are about the same in both tension and compression. Of course, for many brittle materials the characteristic stresses in compression are much greater than in tension.

The solid bodies include bars with axial loads, shafts in torsion, beams in bending, and columns in compression. The relations among the shearing force, bending moment, and external forces are shown in Table 3-1.

Relations among the shearing force, bending moment and external forces　　Table 3-1

	No-external force segment	Uniform-load segment	Concentrated force	Concentrated couple
External forces	$q=0$	$q>0$　　$q<0$	P at C	m at C
Characteristics of Q-diagram	Horizontal straight line; $Q>0$, $Q<0$	Inclined straight line; Increasing function, Decreasing function	Sudden change from the left to the right; $Q_1 \neq Q_2 = P$	Not changed
Characteristics of M-digram	Inclined straight line; M Increasing function, M Decreasing function	Curves; tomb-like, basin-basin	Dog-ear from the left to the right	Sudden change from the left to the right; Opposite to M; $M_1 - M_2 = m$

Words and Expressions

branch　分支，一部分
behavior　性能，状态
strength　强度
bar　杆件
shaft　轴
torsion　扭转
bend　弯曲

column　柱
compression　压缩
beam　梁
rod　杆
request　满足，要求
rigidity　刚度
stability　稳定性

dimension　尺寸
component　成分，组分
relationship　关系
stress　应力
strain　应变
specimen　试件，试样
force　力
elongation　拉长，伸长
measure　测量
cross-section　横截面
plot　绘制
ordinate　纵坐标
proportional　比例的，成比例的
linear　直线的
appreciable　可估计的，可看到的
yielding　屈服
plastic　塑性的
fracture　断裂
lateral　侧面的
contraction　收缩，缩减
neck　颈缩
whereas　而，却，反之
diminish　减少，缩减
withstand　经受，承受
vertical　垂直的，竖立的
transition　转变，过渡
nonlinear　非线性的
ductile　可塑的，延性的
brittle　脆性的
ceramic　陶瓷制品
concrete　混凝土
mechanics of materials　材料力学

applied mechanics　应用力学
solid body　固体，实心体
deformable body　变形体
axial load　轴向荷载
research object　研究对象
theoretical foundation　理论基础
calculation method　计算方法
reasonable shape　合理的形状
proper material　适宜的材料
tensile test　拉伸实验
testing machine　试验机
stress-strain diagram　应力-应变图
typical shape　典型的形状
horizontal axis　水平轴
corresponding stress　对应的应力
proportional limit　比例极限
low-carbon（structural）steel　低碳（结构）钢
yield point　屈服点
strain harden　应变强化
ultimate stress　极限应力
lateral contraction　横向收缩
calculated value　计算值
stress-strain curve　应力-应变曲线
design purpose　设计目的
general characteristic　一般特征
plastic strain　塑性应变
cast iron　铸铁
metallic alloy　合金
characteristic stress　特征应力
proportional limit　比例极限
ultimate stress 极　限应力

Notes

① *by means of* 通过…的方法。
② *in the form of* 以…样的形式。
③ *is placed in* 放置于…，注意用被动语态表达。

④ *subjected to tension* 施加…的力,如构件施加拉力可表达为 the segment is subjected to tension,也是用被动语态表达。

⑤ *dividing… by …*   dividing A by B 译为 A 除以 B,这是常用的表达除法的短语。

⑥ *in this manner*   以这种方式。

⑦ *is said to have become plastic* 是复合谓语;*actually elongate plastically* actually 和 plastically 均修饰动词 elongate。

⑧ *with further elongation the stress increases* 中 with further elongation 是倒装,表示强调。*or ultimate stress* 是同位语。

⑨ 句中 the bar can carry 做 load 的定语;whereas 引导一个状语从句,表示"反之","而","却"等之意。

⑩ 句中 occurring 相当于前面 which occurs,用作定语修饰 strain。

⑪ 句中 analogous to those in tension 做 diagrams 的后置定语。

Translation Examples

[1] The tasks of mechanics of materials: Under the request that the strength, rigidity, stability are satisfied, the necessary theoretical foundation and calculation method are offered for determining reasonable shapes and dimensions, choosing proper materials for the components at the most economic price. The tensile test is the basic test of the mechanics of materials.

材料力学的任务是:在满足强度、刚度、稳定性的要求下,以最经济的代价,为构件确定合理的形状和尺寸,选择适宜的材料,而提供必要的理论基础和计算方法。拉伸实验是材料力学的基本实验。

[2] The relationship between stress and strain in a particular material is determined by means of a tensile test. A specimen of the material, usually in the form of a round bar, is placed in a testing machine and subjected to tension. The force on the bar and the elongation of the bar are measured as the load is increased. The stress in the bar is found by dividing the force by the cross-sectional area, and the strain is found by dividing the elongation by the length along which the elongation occurs. In this manner a complete stress-strain diagram can be obtained for the material.

通过拉伸实验可确定特定材料中应力和应变之间的关系。通常采用圆棒状的材料试样放在试验机中并承受拉伸荷载。随着载荷的逐步增加,测量杆上的力和杆的伸长率。通过将力除以横截面积求出杆中的应力,通过将伸长量除以发生伸长的区域长度求出应变。以这种方式,可以得到完整的材料的应力-应变图。

Reading Comprehension

Ⅰ. Choose the most suitable alternative to complete the following sentences.

1. A specimen of the material, _____ is placed in a testing machine and subjected to tension.

A. sometimes in the form of a hard round bar

B. generally in the form of a round bar

C. always in the form of an unbreakable bar

D. seldom in the form of a brittle bar

2. The force on the bar and the elongation of the bar are measured _____.

A. when the load is added

B. as soon as the load is reduced

C. where the load is increased

D. once the load is given

3. The proportional limit of low-carbon steels is usually between _____.

A. 3000 psi and 36000 psi

B. 30000 psi and 3600 psi

C. 30000 psi and 36000 psi

D. 3600 psi and 36000 psi

4. With an increase in loading, _____, until at point B a considerable elongation begins to occur with no appreciable increase in the tensile force.

A. the strain increases as rapidly as the stress

B. the strain increases as fast as the stress

C. the strain increases more quickly than the stress

D. the strain increases more slowly than the stress

5. The presence of a pronounced yield point is somewhat unique to steel which is _____.

A. used in most parts of the world

B. a commonly used structural metal today

C. used in any construction

D. regarded as the only building material

II. From the list below choose the most appropriate headings for each of the paragraphs in the text, then put the paragraph numbers 3.1~3.6 in the brackets.

A. Further explanation of the strain before fracture occurring ( )

B. A basic concept of the tensile test ( )

C. Introduction of plastic and brittle materials ( )

D. The illustration of the proportions to the stress-strain diagram ( )

E. The relationship between stress and strain ( )

F. Diagrams analogous to various materials in compression ( )

III. Complete the following sentences with the information given in the text.

1. The force on the bar and the elongation of the bar _____ as the load is increased.

2. Thus, with further elongation the stress increases and it reaches its _____ or, at point D.

3. The material actually withstands an increase in _____ to the point of failure.

4. Both _____ and _____ will undergo large strains before failure and are therefore

classified as ductile.

5. Diagrams analogous to those in tension may also _____ for various materials _____, and such characteristic stresses as the proportional limit, _____ and the ultimate stress can _____.

Ⅳ. Choose one word or expression which is the most similar in meaning to the word underlined in the given sentence.

1. He collects <u>specimens</u> of all kinds of rocks and minerals.
   A. samples   B. examples   C. speciosity   D. instance
2. The long war greatly <u>diminished</u> the country's wealth.
   A. narrowed   B. relieved   C. contracted   D. decreased
3. It is the <u>ultimate</u> point of land before the sea begins.
   A. finishing   B. synthetic   C. last   D. eventual
4. The soldiers have to <u>withstand</u> hardships.
   A. confront   B. combat   C. oppose   D. gleam
5. The wages of men averaged $ 54.54, <u>whereas</u> women's wages averaged $ 42.13.
   A. while   B. thus   C. however   D. moreover
6. Match the words in Column A with their corresponding definition or explanations in Column B.

A	B
1. pronounced	a. hard but easily broken
2. stress	b. making longer
3. elongation	c. fact which illustrates or represents a general rule
4. brittle	d. very noticeable
5. ceramics	e. articles made of porcelain, clay, etc.
	f. moving, tending to move, away from the centre or axis
	g. tension, force exerted between two bodies that tough, or between two parts of one body
	h. strong base of a building, usually below ground-level, on which it is built up

Activities—Discussion, Speaking & Writing

Presentation

Group: 5 to 7 members

10 minutes per group (Each member should cover your part at least one or two minutes).

Clearly deliver your points of the following questions to audiences.

NEED practice (individually and together)!!

Gesture and eye contact.

Smile is always KEY!! Cover your nervousness!!

Questions for discussion and presentation

How many experiments have you done in your undergraduate career? Share others with experiments in mechanics of materials or reinforced concrete structure and steel structure.

Writing

Write an article about the experiment you are most interested in.

Further Reading

Reading Material

Mechanical Properties of Ductile and Brittle Materials

The main difference between brittle and ductile materials is that the brittle materials break down after a small deformation, whereas the ductile materials ultimately fail only after considerable deformation. Therefore, the area under the diagram for ductile materials is considerably greater than that for brittle materials.

The amount of work required to crush ductile materials is greater than that required for brittle materials. Therefore, ductile materials are more suitable for structures designed to absorb the maximum possible kinetic energy of impact without failure.

The brittle materials fail easily under impacts just because their specific work of deformation is very small. Due to their small deformation up to stresses close to the ultimate strength, the same brittle materials are sometimes capable of bearing far greater stresses than the ductile materials provided deformation is under the action of a placid, gradually increasing compressive force.

The second distinguishing feature between these materials is that in the initial stages of deformation, the ductile materials may be considered to behave identically under tension and compression. The resistance of an overwhelming majority of the brittle materials to tension is considerably lower than their resistance to compression. This restricts the field of application of brittle materials or requires that special measures to be taken to ensure their safe working under tension as, for example, in reinforcement of concrete elements, working under tension, with steel.

A sharp difference between the behavior of ductile and brittle materials is observed with respect to the so-called local stresses, which are distributed over a comparatively small portion of the cross section of the element but the magnitude of which exceeds the average or nominal value, calculated from common formulas.

Since we do not observe any considerable deformation in brittle materials almost up to the moment of failure, the non-uniform stress distribution shown above remains unchanged under tension as well as compression right until the ultimate strength is reached. Due to this, a weakened bar of brittle material with local stresses will fail or crack at a very lower value of the average normal stress $\sigma = \dfrac{P}{A}$ as compared to a similar bar without local stresses. Thus, we may say that local stresses greatly reduce the strength of brittle materials.

The ductile materials are affected by local stresses to a very lower degree. The role of ductility as regards local stresses is to level them to some extent.

We have given a very simplified picture of the working of a bar with a non-uniform distribution of stresses. Actually levelling out of stresses is hindered not only by strain hardening, but also by the change in the stressed state at the location of stress concentration, its transition from a linear stressed state to a three-dimensional stressed state.

There is one more factor which stipulates the selection of one or another type of material for practical purposes. Often, while assembling a structure, it is necessary to bend or to straighten a bent element. Since the brittle materials are capable of withstanding only very small deformations, such operations on them usually give rise to cracks. The ductile materials capable of taking considerable deformations without rupture, can be bent and straightened without any difficulty.

Thus, brittle materials have poor resistance to tension and impacts, which are very sensitive to local stresses and cannot bear change in the shape of elements made from them.

The ductile materials are free from these drawbacks; therefore ductility is one of the most important and desirable properties in materials.

The points in favour of brittle materials are usually cheaper and often have a high ultimate strength under compression; this property may be utilized for work under placid loading.

Thus, we see that ductile and brittle materials have exceedingly different and contrasting properties as far as their strength under tension and compression is concerned. However, this difference in properties is only relative. A brittle material may acquire the properties of a ductile materials, and vice versa. Both brittleness and ductility depend upon the treatment of the material, stressed state and temperature. Stone, which is conventionally a brittle material under compression, may be made to deform like a ductile material; in some experiments this was achieved by pressing a cylindrical specimen not only at its faces but also on its side surface. On the other hand, mild steel, a conventionally ductile material, may under certain conditions, e. g. low temperature, behave exactly like a brittle material.

Hence the properties "brittleness" and "ductility", which we assign to a material on the basis of compression and tension tests, are related to the materials behavior only at ordinary temperatures and for the given kinds of deformation. In general, a brittle material may change into a ductile material, and vice versa. Hence it would be more precise to speak not of "brittle" and "ductile" materials but of brittle and plastic states of materials.

It must be noted that a comparatively small increase in the ductility of a brittle material (even up to 2% relative elongation before breakdown) enables its use in a number of cases which are otherwise precluded for brittle materials (in machine parts). Therefore, research work on improving the ductility of brittle materials such as concrete and cast iron demands the maximum possible attention.

Translation Examples

[1] The main difference between brittle and ductile materials is that the brittle materials break down after a small deformation, whereas the ductile materials ultimately fail only after considerable deformation.

脆性材料与延性材料的主要区别是脆性材料经过很小变形就能破坏,而延性材料经过明显的变形才能最终导致破坏。

[2] Due to their small deformation up to stresses close to the ultimate strength, the same brittle materials are sometimes capable of bearing far greater stresses than the ductile materials provided deformation is under the action of a placid, gradually increasing compressive force.

由于脆性材料的应力在接近强度极限以前变形很小,只要压力均匀地逐渐增加,有时脆性材料比延性材料能承受更大的应力。

[3] A sharp difference between the behavior of ductile and brittle materials is observed with respect to the so-called local stresses, which are distributed over a comparatively small portion of the cross section of the element but the magnitude of which exceeds the average or nominal value, calculated from common formulas.

我们观察到延性材料和脆性材料性能的一个突出的区别是局部应力分布范围只占构件截面的相当小的一部分,但应力值却超过按通常公式计算出来的平均或名义值。

[4] The ductile materials are affected by local stresses to a very lower degree. The role of ductility as regards local stresses is to level them to some extent.

延性材料受局部应力影响的程度要低得多。局部应力方面的延性作用在某种程度上使局部应力均匀分布。

[5] Actually levelling out of stresses is hindered not only by strain hardening, but also by the change in the stressed state at the location of stress concentration, its transition from a linear stressed state to a three-dimensional stressed state.

确切地讲,应力的均匀分布过程不仅受到应变强化的阻碍,而且也受到应力集中区域应力状态改变的阻碍。这种应力状态的转变是从单项应力状态到三向应力状态变化的。

[6] In general, a brittle material may change into a ductile material, and vice versa.

一般地讲,一种脆性材料可以变成延性材料,反之亦然。

[7] It must be noted that a comparatively small increase in the ductility of a brittle material (even up to 2% relative elongation before breakdown) enables its use in a number of cases which are otherwise precluded for brittle materials (in machine parts).

我们必须注意到脆性材料中延性较小的增加(甚至破坏之前的相对伸长达到2%)就使它可应用在许多原来脆性材料不能应用的地方(在机械零件中)。

Unit 4
Sydney Opera House

Teaching Guidance for Watching, Listening & Reading

Watch videos, pay attention to the *Words and Expressions* and related *sentences* and *paragraphs*.

Some buildings come to *symbolize* a nation, it's the Opera House that instantly puts Australia on the map. Like other *landmark* buildings, it's *enormously ambitious* and uses trailblazing *building techniques*. This incredible building has it shared secrets hiding in its engineering DNA. There are many theories about what inspired the shape of Sydney Opera House roof. Sails? Nun's hat? And armadillo? It just wouldn't be built possible without a *collapsible toy*, *glue for false teeth*, a First World War *gas mask*, ancient Egyptian *woodcraft*, and a *copper bottom sailing ship*.

4.1 A Landmark Building in Sydney

Even if you have never been to an opera in your life, you know this building, the Sydney Opera House. The heart of every great opera house and concert hall obviously is this: the stage, the seats. This is what it's all about. And from here, we really could be anywhere in the world. But it certainly wouldn't say that——about the structure around it as in Fig. 4-1.

Fig. 4-1 Sydney Opera House

It's certainly not just a boring box around stage. In fact, it's one of the most famous buildings in the world. This building has such a famous shape that you could draw on the back of an envelope and almost anyone would recognize it. Completed in 1973, *the distinctive profile* of Sydney Opera House instantly made it an *icon* of Australia. But construction had been troubled. Political running let the *costing overrun* and delay completion by nine years. And the building wouldn't even stand up unless designers overcame formidable *engineering challenges*, with the help from a collapsible toy.

On the top of the building, take a look around by yourself, you suddenly feel you really are standing on a piece of *sculpture*. Just take some breath away. It just does not seem real. There are many theories about what inspired the shape of Sydney Opera House roof. Sails? Nun's hat? And armadillo? Perhaps not. But whatever you see in this is incredible forms. We do know that architect want to create a magical space, where you could leave everyday life behind.

The site itself was *extraordinary*. Previously it was a *tram terminal station*. Danish architect Jorn Utzon, aimed to make the most of its position right on the *waterfront* in the heart of Sydney. This is how it all started in 1956 with Utzon's *rough sketch* for a competition for new opera house, as shown in Fig. 4-2.

According to one version of the story, Jorn Utzon's winning design with these *spectacular shapes* was not even on the short list. It was picked out of the *reject bin* by one of the judges.

But engineer Aoveor, who is another hero role of the story, felt he could not build Utzon's winning design. The problem was lots of big *curved shapes*. And Utzon did not want *columns* to hold up, either. *Steel* would be obvious solution. It was easy to work and strong enough to hold complex shapes, best of all it would be *affordable*.

Fig. 4 – 2 Design sketch of the opera house

But Utzon didn't think about the price or making construction simplified. His plan was for a *vast magical space*, a huge sculpture. He was an architect and he wanted to use the material as-concrete. It was up to engineers to make it happen.

4.2 Inspiration I: A Collapsible Toy & the Technique of Post-tensioning

Oryana's answer was to make a *frame* for each sail with huge *hollow concrete ribs*. But it had never done before on the scale required in Sydney. The biggest ribs would be fully 55 meters high. They were too big and heavy to *mold* in one piece. The principle behind the solution was child's play. And it was our first *connection*, a *puppet*.

This is the key, not this particular *giraffe* in Fig. 4 – 3 but the principle behind it. Because toys like this are made of individual *segments*, held together with *cord*. When you press the button on the bottom, the cord slack and whole thing *collapses*. But let it go again, as the cord tied up, the whole thing keeps the shape, as shown in Fig. 4 – 3. And that is the principle they used when they built the Sydney Opera House.

Oryana decided to make the *huge ribs* out of *segments* which could be put together to make

(a)　　　　　　　　　　　　　　　　(b)

Fig. 4 – 3 The display process of Giraffe Doll Principle

the right shape, just like that children collapsible toy. The technique is called *post-tensioning*, it can also *strengthen* the concrete. It was a device by a French engineer in the early of 1920s century to make big spans in bridges.

Back to the workshop, Oryana engineer Ed Clark showed how it worked.

Ed clark: The Sydney Opera House was made of *precast segments* concrete, we got *precast* segments, not concrete here, but *polystyrene*. So we can stack them up and build themselves an arch.

The host: So we are going to recreate it here on the 1/10 *scale*, it's quite big, and it can be built as 6.5 meters high. The Sydney Opera House is about 60 meters high.

We started by building two stacks of *block*, the post-tensioning came later when you *thread* all the blocks together. There are actually concrete, which are just very strong. Soon, we had two *curved* stacks of blocks resting on the *steel supporting frame*. Next, we needed to put in the post-tensioning *cables*, this sort of puppet's strings will hold the arch up.

Ed clark: These cables were like washing lines that didn't use in Sydney House. We were using the same *principle*, which was just in the smaller scale. Let's thread in the holes in the plate there.

Once the cables have been threaded, we needed to tighten or post-tension them. Because *it was* still the steel frame that hold the arch up, as in Fig. 4-4.

(a)
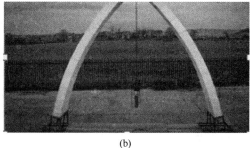
(b)

Fig. 4-4 Pointed Arch Tension Test

Ed was happy that the cable had been evenly tensioned, we were ready to remove the *steel frame*. After removing the steel frame, the arch worked and it stood up. Ed then wanted to *tension* our puppet strings even more to increase the *strength* of the *arch*.

The host: Is it possible that as we tied it over, the whole thing fell over?

Ed Clark: That's a danger if we over tied it. The polystyrene is of a certain *compressive strength*. The more we tied them, the more *bending resistance* they will have. But if we over tied here, the polystyrene could *crash* as the same with the *concrete*.

Finally, we were ready to test our post-tension *polystyrene arch*.

Ed Clark: I think we should hang a test load from the top to see how it is going.

The first load was 40 kilograms weigh.

Ed Clark: We could hear a little bit crunch. It seemed ok, we can add another 40 kilograms' load.

It was just made of polystyrene and a couple of small *wires*, which was 7 meters in the air. But Ed was confident that polystyrene could take a lot of more. Another 70 kilograms were loaded, bringing it to total 150 kilograms, and still the arch stood strong. There was only one way to go, taking it to the limit of 240 kilograms, which are half of a horse.

The host: It was just a big stack of *polystyrene blocks*, post-tensioned with *narrow wire*.

Ed Clark: It carried about a quarter of a ton.

The host: I couldn't believe it could hold its own shape, let alone support another a quarter of a ton. And seeing post-tension *behave* like that, it *strengthened* the link between the structure and the giraffe here. Because as you can see, it's just a series of sections held together by the tension in the cable running through the center on it. With no tension, they were just part of different bits. When put tension in, it stood up. And that gave me an idea for one last *illustration*, what if we take the tension out of that system…

Ed Clark: We could do that (taking the tension out of the polystyrene).

(All the cables were cut at the same time and the structure collapsed).

The host: We could safely conclude that it was the tension in these cables, holding up the whole structure. Post-tensioning obviously worked.

4.3　Inspiration Ⅱ: A Peel of Fruit (Orange) & the Magical Space

Making a huge concrete puppet allowed the Sydney engineers to build a complex shape. But it left them with a different problem. The secret of the cheap production is to make the same shape over and over again, rather than products with different shapes and sizes like the Sydney Opera House. Each Sydney sail *rib* curved differently, and we needed different shapes of *moulds*, which would be expensive.

Utzon found the answers to building this magical space by this orange. It wasn't a magical orange with answer in it. It was a shape, because as he peeled the fruit, he realized he got it. Each of the sails was a different shape, but they could all come from one curve.

It's incredible, but if you took a huge orange with 150 meters across, you could make the Sydney Opera House's sails out of pieces of the peel. Each sail is the different size piece of the peel. Because they are from the same orange, all the sails curve in the same way (Fig. 4-5a). Each of the different sail ribs could be made of the same curved segments (Fig. 4-5b). A few were for the small ribs while more for the big ones. Production could be simplified.

Sydney Opera House's roof construction began in 1963, but already the project was under strain, with *political opponents* slandering Utzon with *over soaring costs*. The project called for the largest *crane* in the world to lift the concrete ribs. A special kind of fearless workers manually moved each *segment* into position. Sadly one builder was killed. But even so, in 14 years of *construction*, often with no safety harness, it was still an *astonishing record*. They locked down each

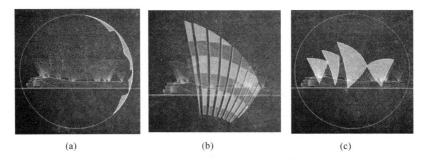

Fig. 4 – 5 Different Orange Peels Create Curved Outline

rib with huge *steel cables*, their own version of the *puppet strings*.

The host: These great concrete blocks were equivalent to the segments in the little giraffe toy. So the ribs were tensioned vertically to stop them *slipping*, and *horizontally* to stop them *falling apart*. I haven't yet found a button like the giraffe toy you can press to release the tension.

And because *squeezing* the concrete also made it stronger, it allowed thinner and more *delicate* looking sails, perfect for Utzon's design.

4.4 Inspiration Ⅲ: A Glue for False Teeth & the Precast Segments Concrete

Post-tensioning in the sails made engineers didn't have to build huge *columns* inside the Opera House, which would have been inconvenient when you try to watch a ample lady or man singing.

But not even the tension steel was enough to lock everything in place. Heat causes the concrete to expand, and it contracts when they cool. Relentless *swelling* and *shrinking* could force the concrete *segments* apart, not good for water tight *joints*. Worse, it put not even stress on the post-tension structure itself, which could be a headache for opera goers.

The host: The *tiny movement* in the segments of the roof could have be *fatal*. One slack, and this whole thing would've collapsed like a pack of cards. So it has to be well glued.

Fig. 4 – 6 False Teeth

To insure everything *stay in place*, the engineers needed a very special kind of *glue*. They have to look to *dentistry*. This false teeth in Fig. 4 – 6 were made more than 2000 years ago. A *bulky metal* brace held them together. There was no good glue to hold teeth in dentist or Opera House up until 1936, when Pear Casden produced first epoxy glue and revolutionized the World's false teeth.

The host: Casden's *epoxy resin* created strong *bonds* perfectly keeping false teeth. But perhaps more importantly it also means the Opera House can stay up.

Before Casden's dental revolution, glues needed *surfaces* to be absorbed, like wood. In the 1930s, a plane like a mosquito bomber could be stacked together with glue only because it had a wooden airframe. The glue permeated in, making a strong joint. As *planes* involved, the designers turned to metal, which just like the concrete of the Opera House in which glue wasn't absorbed. And the pioneering sea HORNET entered in the early 1940s. Sticking the metal in this plane needed a new kind of glue that formed a strong bond with the *molecule* on the *surface*. This family of new glues that formed new molecule bond included Casden's epoxy.

The host: Back to my workshop, I have deviced a sticky experiment to compare this kind of Casden's device with the old glue. Adhesive expert John Bishop helps me with the epoxy first.

The host: What should I do?

John Bishop: *Squeeze* the *trigger* and put a bit up and down.

When the glue had set, it was time for little test.

It was going to go head to head with the wooden glue used on the mosquito airplane. Could either of them stick concrete or even metal?

The first specimen was the glue to hold airplanes made by wood.

The task was to assemble the specimen on the car, which would be lifted to test the glue's viscosity, as shown in Fig. 4 – 7.

When lifting the car, the glue didn't work at all.

The host: You know, frankly it's not supposed to work at metal. What we can learn from its interface?

John Bishop: It had to penetrate in the surface, but it obviously hasn't done. So it wouldn't work on metal. You can peel this glue off quite easily. So that's why it didn't stick, that's why you don't apply on the airplane of metal, which was not stuck with the sort of glue.

(a)   (b)

Fig. 4 – 7   Viscosity Comparison Test

And wood glue like that wouldn't work on concrete either. It simply couldn't have stuck the sails together at the Opera House.

Test two, glue like Casden's dentures epoxy was *apparently* so stronger that much less of the area with glue should hold, even on metal.

The host: So that's the quarter of the area with the stick that applied. (The car was lifted and the glue worked, as shown in Fig. 4 – 7b). Oh, that's clearly quite happier.

When the car rocked, it dramatically increased the load on the metal joint, but this glue still worked well.

The host: Just how strong is that stuff?

John Bishop: About *4 tons per spare inch*, 4 tons in shear.

The same epoxy that keeps teeth rooted in dentures, sticks the ribs of the Opera House together, and guarantees that they won't slide against each other when *heating swells* and *cooling shrinks* the concrete. The *skeleton* was complete, now it needed a beautiful skin.

There are more than a million *tiles* on this roof. And Utzon asked for two types, *matte* and *glows*, and they were laid in intricate patterns, and then fixd in lids, made of chicken wire and concrete, which were then fixed on the concrete ribs and the roof of the Sydney Opera House was complete.

But while more fearless workers laid the tile lids on the building, behind the scene, the project was *interminable*. The architect himself wasn't around to witness the completion of his *iconic* roof. In a furious dispute of rising building costs, Utzon walked away from the project in 1966. But construction continued. Oryana engineers moved on to the next challenge.

4.5 Inspiration Ⅳ: Gas Mask in First World War & the Large Glass Windows

The next challenge was windows, Utzon wanted concert goers to be able to see and enjoy the building's spectacular location, right in the heart of the city on the harbor front. But there were 3 problems: The size of the windows, the shape of the windows, and the fact that the risk of killing people if you use ordinary glass.

The designers wanted massive windows to allow the light to flood in, but big pieces of *ordinary glass* were weak, relatively only small shocks could shatter them. Nobody wanted 4 meter-high windows falling onto the Opera goers. It took a deadly threat back in the First World War that poison gas pointed to the answer. Without *gas mask*, it just wouldn't have been possible to build Utzon's Opera House safely. The same invention that prevented glass eyepieces from shattering and blinding soldiers' eyes would come highly when making Utzon's ambition a reality. To see why normal glass was not such a good idea either for gas masks or the Opera House, the host visited one of the biggest *glass makers* in the world. *Glass expert* Susan Lambeth in Saint-Gobain prepared a big sheet of standard glass.

Susan Lambeth: Ok, well, this is a piece of 10 millimeters-thick *float glass*, it is what we make here everyday.

The host: What might I use a piece of glass like this for?

Susan Lambeth: It could be typically used in windows and for sides and some roof and things like that at the size and thickness.

The host: So this is just a big piece of *standard glass* as it comes out from *manufacturing* blocks right over there. It is a basic *sample* from here.

This is good glass as used in ordinary buildings all over the world everyday. We are just going to smash, namely test it.

In test, the watermelons were used as Opera House goers. A piece of tile was the essential *equipment*.

The host: I'm going to drop a tile with the same size as on the Opera House from more than 10 meters. In reality, the Opera House is 55 meters high.

The tile was hit on the glass. Not a *scratch*, it proved to be a good glass. But of course, everything had a limit.

Then the test was for a bigger one. The brick would be dropped from more than 10 meters.

The host: I've done a math, this *brick* from this height of 11.5 meters onto the glass, had the same *impact* as one of those roof tiles falling off the tallest sail at the Sydney Opera House onto a piece of glass. You see, it's *sophisticated*.

The glass was in tatters. Things were not good for "watermelon audiences", as shown in Fig. 4 - 8. It was *substancial* pieces of glass.

(a)　　　　　　　　　　　　(b)

Fig. 4 - 8　Test of Ordinary Glass

That glass is perfect for everyday use, but at Sydney Opera House this would've been in *carnage*. There was a solution, as used in World War I gas masks, *laminated glass*. Amazingly, there is only one tiny little difference from the *standard glass*. Back to the factory the glass starts its life in fire *furnace*, where the raw ingredience was mixed and baked. 800 minutes later, the Continued sheet is cut into *panes*. The secret is to make these panes into a special sandwich. A thin film of plastic polyvinyl butyral or PVB, is sandwiched between the two panes of standard glass. The result is the laminated glass.

The tiny bit stuff really makes a big difference. And this whole process that making glass sandwich was discovered by accident.

Reback to a French laboratory in the early 1920s, a clumsy scientist knocked over an empty chemical flask. Magically it didn't shatter. Nobody has cleaned the glass, the broken flask held its shape because of an almost invisible film of sticky nitrocellulose. The idea of laminated glass was born. Cut to the First World War, troops needed protection from deadly poison gas attack, so the gas mask was born. But as we've seen, bombs and fragments made standard glasses of eyepieces create this kind of splinters which can split watermelons. *Laminated glass* saw action for the First World War. But is it good enough for Opera House's windows?

The laminated glass was tested to suffer the brick dropping down from 11.5 meters high, which was equivalent to a tile falling from 55 meters, just the height of the top sails. As a result, it didn't even scratch. The brick just bounced off, as shown in Fig. 4-9 (a).

(a) Hit by a brick   (b) Hit by a bouling ball

Fig. 4-9 Laminated Glass Cracked But Not Broken

More than 7 kilograms, a bowling ball was only used to test the performance of the glass in this experiment though it rarely drops from the sky. The brick that ruined the same piece of that standard glass just glanced off the laminated glass. Under the boiling ball's hit, even though it was broken, it was still stuck on the plastic film, as in Fig. 4-9 (b).

So thanks to the laminated glass, the safety for "watermelons" were ensured.

Laminated glass was the solution to solve the problem of the *trench warfare* in the World War I, and it offered the answer to the *engineers* at Sydney. Now concert goers can hold a glass in their hands, without fearing glass falling on their heads. The structure of the Opera House was complete. It looks good, and engineers have to make sure it sounds good too.

4.6 Inspiration Ⅴ: Egyptian Pharaoh's Chest & the Complex Shape Inside the Concert Hall

Concrete was never suitable for an actual concert hall. The concrete shapes and material itself were just completely unsuitable for *sound transmission*. Inside the shells, there's *tall pointed space*, where the voice that even strongest of the singer would've disappeared.

Even worse, the concrete is *crustily harsh*. It *reflects* sound just like mirror reflects light.

The designers needed to create an interior with completely different shape and out of completely different material, something like wood.

Its shape had to be curved to bounce sound back down. But they needed a special kind of wood. Ordinary wood wasn't light or strong enough to do the job without huge frame to hold in place. An ancient pharaoh's mummy pointed to the solution. Tutankhamen's tomb was discovered in 1922, along with something much more important for the Sydney Opera House. Among priceless ancient Egyptian antiquities, genius wooden chests were discovered. Despite the fertility brought by the Nile, quality wood was in short supply. To get around it, ancient Egyptian carpenters faked the appearance of fine wood. Tutankhamen's chests were made of layers of poor wood faced with the thin lamina of rare wood.

In other words, it was *plywood*. It probably comes a bit of surprise to someone, because plywood frankly has something of image problem. But it can be used for *furniture*.

Utzon believed that plywood was good for more than just furniture. He thought it could work on a huge scale, to create the perfect interior for the Opera House. The key to molding in the complex shapes inside the concert hall was the way plywood is *constructed*.

Wood is much stronger in one *direction* than another. It is easy to *split* along its *grain*, naturally going that way. The key to plywood strength is to use different layers of wood with their grain in opposing directions. And as the result of that process, a given thickness of the plywood is stronger than the same thickness of plain wood. It needs less material to do the same job, which is lighter. Plywood might look ordinary, but it got hidden strength.

The host: I am going to try to make something that's light and strong. And I'm going to think like Egyptians and make it elegant as well (in Fig. 4 – 10a).

The bridge in Fig. 4 – 10 (b) has been structured entirely from the plywood, by layering and laminating different sheets to form that graceful curve. But plywood isn't just about making beautiful shapes, it's also surprisingly strong.

The host: The bike weighs 50 kilos and I'm another 69, 70 or 71 kilos. As I ride on the bridge, the wood bends, but it bounces back as in Fig. 4 – 10 (b).

(a) Test for plywood (b) Riding a bike to test the strength of plywood

Fig. 4 – 10 Plywood Test

The interior of the Opera House is like a separate building inside the concrete shell. You can get places up in the roof, where you can see you are between the two structures.

The host: *Stagehands* call this the *bomb* bay which is above the *odeum*. And that means that I'm sandwiched between the roof of the plywood shell and the concrete shell up my head.

Plywood is good for the *interior* because its natural *acoustics* is better than concrete. And being light and strong, it could make a separate shell inside the concrete sails. Best of all, Tutankhamen's craftsman knew it can be shaped. The right curve means it reflects sounds back down. So the audience can hear what they pay for.

4.7 Inspiration Ⅵ: A Copper Bottom Sailing Ship & the Air-conditioning System

The host: So that's how they built the Opera House outside and in, except there's just one more thing. It's a beautiful *sculpture* thing, and *spectacular* sighting. And it's a place where concert goers will come to escape from reality. But how can they make sure they're comfortable while they are enjoying the music? The answer lies down there.

All any given night, the Opera House can hold not just opera but ballet, drama, and concerts of all kinds of music. Every morning, the theaters come alive. As *technicians* prepare the stages for the shows, they check sounds and rig lights. But a perfect *performance* requires something else.

It's a warm stuffy evening on the Sydney waterfront. People can wait months for tickets. And it ought to be a beautiful night act.

You come into this *extraordinary* building, the excitements into this room and around it is palpable, you are full of expectation. You take your seat. And then, no matter how hot the performance, if there isn't good air conditioning you're going to be in misery. A room full of thousands of people heats up very quickly. Its outline wouldn't look at the same with a *chimney*, a fan hanging around it, or a cooling tower poking out of the top, which large *air conditioning system* usually needs. So they needed a different solution. Thanks to a copper bottom sailing ship, the answer was all around them. All air conditioning systems need somewhere to get rid of the heat. If they can't go up with the upper chimney, where does it go? Facility manager Bob Muffins led the way to the underneath of the Opera House, to show where they found an inexhaustible supply of something to take the heat away from the Opera House. It's the water from Sydney harbor.

Bob Muffins: Basically, it is the sea water intake tunnel. It's about 1 meter high, it allows the water into the building. And air conditioning process starts from there.

Inside the Opera House, the sea water and fresh water are in the *vessels*, both in its independent circuits with each other.

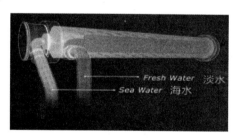

Fig. 4 – 11   Water Circulation System

Inside this large vessel (Fig. 4 – 11), there are two sets of pipes carrying sea water and fresh water in completely separate *circuits*. Heat from the air conditioning system is brought out by the *fresh water*, and is transferred to the cooler sea water. They never actually mix. The *sea water* comes in on this side, cools down the interior of the vessels which is the fresh water and goes out on the other side out to the ocean. It is not a long stay for the sea water. But during that short visit, sea water can make horrible damage to the Opera House's expensive pipe work. It's really *corrosive* stuff, which brings us to sailing ships. To protect 19th century wooden ships from wood borers' impacts, the British navy *shielded* the holes with copper. When they added copper, they found the iron nails holding the boat together rusted. Fascinatingly it was all to do with electricity, as corrosion expert Robin Oakley shows.

Robin Oakley: This is a piece of *copper*, and this is a piece of *zinc*. So they are different metals. Putting together as they're dry, there's no *electricity* flow between them. However, if I put just a bit salty water between them, namely sandwiched single salty water between different metals, we have electricity flow now.

The host: So it's the sea water that makes it possible, because of effective conductance.

Connecting two metals in salt water makes a *primitive battery*. The bad news for the ships and Opera House is it also makes corrosion.

The host: I'm going to show how some metals are pro to corrode while others resist strongly.

Robin Oakley: The most resistance metal is gold. They don't corrode at all. Because gold is called noble metal, the metals that are less noble will corrode. Copper is quite *noble*. And we got a more active metal, a painted steel plate with a suitable image of the Opera House, as in Fig. 4 – 12. The image of the Opera House is not painted. So what then happens is *electric current* caused by the two metals, will focus into those bare metal parts.

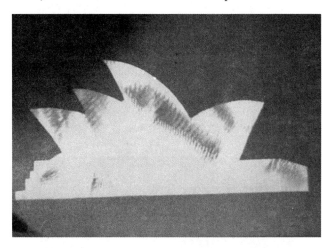

Fingure 4 – 12 The original experimental plate with Sydney Opera House image

The next step in reality normally takes years, and we would speed this copper a bit, with a car battery, fast forward 1000 times.

Robin Oakley: We're going to accelerate the process in its nature direction, but just harder and faster. It's relatively *low voltages* with little danger. (The experiment started) The copper plate didn't corrode, that's not *corrosion*. *Corrosion* is happening in the steel plate, but it's slow. That is gas, we actually break the water apart, and it is given hydrogen gas, which is spectacular as you can see it.

24 hours of furious bubbling, produced 3 years work of *swamp*.

Robin Oakley: It is all basically *rust*. The steel has been corroded. And now we'll find out how much the accelerated *corrosion* is done to the plate.

The host: Wow! That came through.

Robin Oakley: Yeah, that corrosion has eaten away this "Sydney Opera House" pattern at the middle of the plate, as in Fig. 4 – 13. So this exampled how mixed metals in a conductive liquid like sea water, produced the electricity, which can be destructive to cause corrosion.

The steel was eaten away because it's more reactive, which is less noble than copper. If you want to protect the steel, a metal that even less noble should be sacrificed. We got another tank that fixed for 24 hours with the same copper and steel, only difference was to put some sacrificial zinc.

Robin Oakley: What done in this device was the same set-up, with a zinc block electrically connected to this steel panel.

The host: So you introduced a third metal in there.

Robin Oakley: Yes, let's see what has done to this steel plate.

The host: There is nothing on that plate, no influence at all, as in Fig 4 – 14.

Figure 4 – 13　The corrosion has "eaten away" the "Sydney Opera House"

Figure 4 – 14　The steel panel protected by the zinc block

Robin Oakley: The zinc here should have been corroding preferentially, being sacrificial to the steel. Actually this wasn't the original zinc block, it has eaten away already.

The host: So these whole things become a system that you introduce zinc into. The zinc is really the most corroded than anything else in that system.

To stop those *iron nails* in copper bottom sailing ships corroding, they protect them with less noble metal. So Humphrey David attached zinc to the hole, the zinc was sacrificed and renewed as necessary. And at the Sydney Opera House, they protect that steel pipes from salty water by putting replaceable sacrificial zinc in the sea water circuit.

Billions liters of sea water have passed through the Sydney Opera House in 35 years of operation, but its air conditioning system is still uncorroded. Sydney Opera goers can bath in the music rather than their own sweat, thanks to some bits of zinc on a copper bottom sailing ship. Best of all, Utzon's magical space keeps its distinctive profile.

Whether those are sails or nun's hat or sea shells, that beautiful building connects Sydney with this beautiful harbor. Utzon gave Austrialia a design for new opera house, but without Aoveor and his structural designers, it wouldn't be built as one of most distinctive pieces of architecture anywhere in the world—— Austrialia's *landmark building*.

And Utzon and Aoveor couldn't have done it, without a collapsible toy, a clumsy scientist, false teeth, a pharaoh's chest, and a copper bottom sailing ship.

Words and Expressions

symbolize	象征，标志，用符号做代表
landmark	地标，陆标，里程碑
woodcraft	木材加工术，木工技术
icon	偶像，象征，图标
sculpture	雕塑，雕刻品
extraordinary	非凡的，特别的
waterfront	滨水区，码头区
column	柱
steel	钢，钢铁
affordable	负担得起的
frame	构架，体系，结构
mold	塑造，用模子制作
connection	连接，关联
puppet	木偶
giraffe	长颈鹿
segment	部分，段，片
cord	细绳，索，带
collapse	倒塌，坍塌
strengthen	加强，巩固
polystyrene	聚苯乙烯
scale	大小，规模，比例，比率
block	大块
thread	穿，穿过，穿行
curve	曲线，弧线
cable	钢索，绳缆，电缆
principle	原理，原则
tension	拉紧，绷紧
strength	强度
arch	拱，拱门，拱形结构
crash	坠毁，坠落
wire	电线，导线
behave	表现
section	部分

illustration	例证，实证
rib	肋，拱肋
mould	模子，模具
crane	起重机，吊车
construction	建造，构造
slip	滑落，滑脱
horizontally	水平地，平行地
squeeze	用力挤压，榨取
delicate	精致的，精细的
swell	膨胀，增大
shrink	收缩
joint	连接点，接合处
fatal	致命的，灾难性的
glue	胶，胶水
dentistry	牙科
bond	结合力，纽带，联系
surface	外部，表面
plane	飞机
molecule	分子
trigger	扳机，启动装置
apparently	显而易见地
skeleton	骨骼，骨架
tile	瓦片，瓷砖
matte	不光滑的
glow	色彩强烈，绚丽夺目
interminable	冗长的，无止境的
iconic	图标的，符号的
manufacturing	制造，生产
sample	样品，标本
equipment	设备，器械
scratch	划痕，刮痕
brick	砖，砖块
impact	效果，影响

sophisticated　先进的，精密的，复杂的
substantial　实质的
carnage　大屠杀
furnace　火炉，熔炉
pane　框格玻璃
stuff　东西，物品
engineer　工程师
reflect　反射
plywood　胶合板
furniture　家具
construct　构成
direction　路线，方向
split　裂开，分裂
grain　纹理，纹路
stagehand　舞台工作人员
bomb　炸弹，爆炸装置
odeum　音乐厅
interior　里面，内部的
acoustics　传声效果
spectacular　壮观的
technician　技术员
performance　表演，表现
chimney　烟囱，烟道
vessel　容器，器皿
circuit　环形线路，环形
corrosive　腐蚀性的
shield　遮蔽，保护
copper　铜
zinc　锌
electricity　电，电能
noble　贵重的，宏伟的，壮观的
corrosion　腐蚀物
swamp　沼泽，湿地
rust　锈，铁锈
enormously ambitious　雄心勃勃的
building technique　建筑技术
collapsible toy　可折叠玩具
glue for false teeth　假牙胶水
gas mask　防毒面具

copper bottom sailing ship　铜底帆船
the distinctive profile　独特的轮廓
costing overrun　成本超支
engineering challenge　工程挑战
tram terminal station　有轨电车终点站
rough sketch　草图
spectacular shape　壮观的形状
reject bin　废品箱
curved shape　曲线形状
vast magical space　巨形魔法空间
hollow concrete rib　混凝土空心肋（板）
huge rib　巨大的肋（板）
post-tensioning　后张法
precast segment　预制段
steel supporting frame　钢支撑框架
steel frame　钢框架
compressive strength　抗压强度
bending resistance　抗弯性
polystyrene arch　聚苯乙烯（尼龙）拱
polystyrene block　聚苯乙烯（尼龙）块
narrow wire　窄线
political opponent　政治对手
over soaring cost　成本飞涨
astonishing record　惊人的记录
steel cable　钢缆
puppet string　木偶弦
fall apart　四分五裂
tiny movement　细微的移动
stay in place　留在原地
bulky metal　大块金属
epoxy resin　环氧树脂
4 tons per spare inch　每平方英寸4吨
heating swell　加热膨胀
cooling shrink　冷却收缩
ordinary glass　普通玻璃
glass maker　玻璃厂商
glass expert　玻璃专家
float glass　浮法玻璃
standard glass　标准玻璃

laminated glass　层压玻璃，胶合玻璃
trench warfare　壕沟战
sound transmission　声音传输
tall pointed space　高耸空间
crustily harsh　残酷的
air conditioning system　空调系统
fresh water　淡水
sea water　海水

primitive battery　原始电池
electric current　电流
low voltage　低电压
iron nail　铁钉
zinc block　锌块
control the corrosion　控制腐蚀
landmark building　地标建筑,标志性建筑

Translation Examples

[1] Oryana's answer was to make a frame for each sail with huge hollow concrete ribs. But it had never done before on the scale required in Sydney. The biggest ribs would be fully 55 meters high. They were too big and heavy to mold in one piece.

奥雅纳的办法是用巨大的中空混凝土拱肋来打造支撑每幅船帆的框架，但所需的规模在悉尼实属空前。最大的拱肋足足有55米高，无论体积和重量都无法支模制成一整块。

[2] Oryana decided to make the huge ribs out of segments which could be put together to make the right shape, just like that children collapsible toy. The technique is called post-tensioning, it can also strengthen the concrete.

奥雅纳决定用个别组件构成巨大的拱肋，把组件串接起来形成应有的形状，就像儿童的可折叠玩具，这种工法称为后张法，也能够增强混凝土。

[3] The Sydney Opera House is made of precast segments concrete, we got precast segments, not concrete here, but polystyrene. So we can stack them up and build themselves an arch.

悉尼歌剧院是以预制混凝土构件建成的，我们这里也有预制构件，不过这儿它们不是混凝土而是聚苯乙烯，我们可以把它们堆起来建成一个拱。

[4] We started by building two stacks of block, the post-tensioning came later when you thread all the blocks together.

我们开始建造出两叠砌块，稍后再使用后张法把所有砌块串联起来。

[5] It's incredible, if you took a huge orange 150 meters across, you could make the Sydney Opera House's sails out of the pieces of the peel. Each sail is the different size piece of the peel. Because they are from the same orange, all the sails curve in the same way.

听起来不可思议，但如果拿一个直径150米的大橘子，就能用剥下的果皮组成悉尼歌剧院的船帆。每只船帆就是一块尺寸不同的橘子皮，因为是同一个橘子，所以每只船帆的曲度完全相同。

[6] It took a deadly threat back into the First World War that poison gas pointed to the answer. Without gas mask, it just wouldn't have been possible to build Utzon's Opera House safely. The same invention that prevented glass eyepieces from shattering and blinding soldiers' eyes would come highly when making Utzon's ambition a reality.

第一次世界大战中的致命威胁毒气指向了问题的答案，如果没有防毒面具的灵感根本不可能打造出伍重安全的歌剧院。原本是用来防止玻璃目镜碎裂后刺瞎士兵双眼的发明在实现伍重的设计构想时发挥了作用。

［7］I've done a math, this brick from this height of 11.5 meters on to the glass, had the same impact as one of those roof tiles falling off the tallest sail at the Sydney Opera House onto a piece of glass.

我通过数学计算获得，这个砖块从 11.5m 的高度落到这块玻璃上同一块悉尼歌剧院船帆屋顶最高处的瓷砖落下去打到玻璃窗上的效果是相同的。

［8］The designers needed to create an interior with completely different shape and out of completely different material, something like wood. Its shape had to be curved to bounce sound back down. But they needed a special kind of wood.

设计师打造的室内空间形状必须完全不同，采用截然不同的建材，就像木材。空间必须呈弧形，他们需要一种独特的木材打造不同弧度的空间造型，能把声音往下回弹。

［9］The key to plywood strength is to use different layers of wood with their grain in opposing directions. And as the result of that process, a given thickness of the plywood is stronger than the same thickness of plain wood.

胶合板强度的关键在于利用不同纹理的木材来顺应不同的受力方向。这样处理的结果，使相同厚度的胶合板比纯木板强度高。

Activities—Discussion, Speaking & Writing

Presentation

  Group: 5 to 7 members
  10 minutes per group (Each member should cover your part at least one or two minutes).
  Clearly deliver your points of the following questions to audiences.
  NEED practice (individually and together)!!
  Gesture and eye contact.
  Smile is always KEY!! Cover your nervousness!!

Questions for discussion and presentation

  Some buildings come to symbolize a nation, it's the Opera House that instantly puts Australia on the map. Like a landmark building, it's enormously ambitious and uses trailblazing building techniques. There are many theories about what inspired the shape of Sydney Opera House roof. Sails? Nun's hat? And armadillo? How many challenges and connections were on Sydney Opera House roof? And what are they?

Writing

  Read Sydney Opera House, write a short essay independently which contains at least the following information.

悉 尼 歌 剧 院

1 建筑特征

悉尼歌剧院是 20 世纪最具特色的建筑之一，也是世界著名的表演艺术中心、悉尼市的标志性建筑。这座综合性的艺术中心，在现代建筑史上被认为是巨型雕塑式的典型作品，在 2007 年 6 月 28 日被联合国教科文组织列入《世界文化遗产名录》。歌剧院规模：悉尼歌剧院整个建筑占地 1.84 公顷，长 183 米，宽 118 米，高 67 米，相当于 20 层楼的高度。悉尼歌剧院坐落在悉尼港湾，三面临水，环境开阔，以特色的建筑设计闻名于世，悉尼歌剧院的外形犹如即将乘风出海的白色风帆，与周围景色相映成趣。

2 外观结构

悉尼歌剧院的外观为三组巨大的壳片，耸立在南北长 186 米、东西最宽处为 97 米的现浇钢筋混凝土结构的基座上。歌剧院整个分为三个部分：歌剧厅、音乐厅和贝尼朗餐厅。歌剧厅、音乐厅及休息厅并排而立，建在巨型花岗石基座上，各由 4 块雄伟的大壳顶组成。高低不一的尖顶壳，外表用白格子釉瓷铺盖，在阳光照映下，远远望去，既像竖立着的贝壳，又像两艘巨型白色帆船，飘扬在蔚蓝色的海面上，故有"船帆屋顶剧院"之称。那贝壳形尖屋顶，是由 2194 块、每块重 15.3 吨的弯曲形混凝土预制件，用钢缆拉紧拼成的，外表覆盖着 105 万块白色或奶油色的瓷砖。据设计者晚年时说，他当年的创意其实是来源于橙子，正是那些剥去了一半皮的橙子启发了他。而这一创意来源也由此刻成小型的模型放在悉尼歌剧院前，供游人们观赏这一平凡事物所引起的伟大构想。

3 建造历程

设计：1956 年，丹麦 37 岁的年轻建筑设计师约翰·伍重看到了澳大利亚政府向海外征集悉尼歌剧院设计方案的广告。1957 年冬天，约翰·伍重的方案击败所有 231 个竞争对手，获得第一名，人们都为其独具匠心的构思和独树一帜的设计而折服了。但其实约翰·伍重的方案最初很早就遭到了淘汰，被大多数评委否定而出局。后来评选团专家之一，芬兰籍美国建筑师埃洛·沙里宁来悉尼从废纸堆中重新翻出，他看到这个方案后立刻欣喜若狂，并力排众议，在评委间进行了积极有效的游说工作，最终确立了其优胜地位。歌剧院的建造计划一共有三个阶段。

阶段一：建造矮墙（1959 年~1963 年）

于 1958 年 12 月 5 日开始，建筑公司为 Civil & Civic，奥雅纳工程顾问公司的工程师们则负责监督和指导。政府出于对资金和公众舆论的担心力求工程尽快开展。然而约翰·伍重的最终设计却仍未完成。1961 年 1 月 23 日，工程已比预计延后了 47 天，这主要是因为遇到了一些没有预料到的困难（包括天气、没有预料到的雨水改道、工程在正确的结构图准备好之前就已开始、合同文件的改变）。矮墙的工程最终于 1962 年 8 月 31 日完成。迫使工程尽快开展的行为导致后来产生了一些显而易见的问题和这样的一个事实：矮墙的强度并不能够支撑它的屋顶结构，因此必须要重建。

阶段二：建造外部的"壳"结构（1963 年~1967 年）

在最初的歌剧院设计竞赛中，这些壳并没有几何学上的定义，但在设计过程的开始阶段，这些"壳"被定义为由一系列的混凝土构件组成的排架支撑起来的抛物线。然而，奥雅纳工程顾问公司的工程师们找不到建造这些"壳"的方法。使用现场浇筑的混凝土来建造的计划由于造价高昂而遭到了否决，因为屋顶的结构不同，要求有不同的模具，最终会导致造价高昂。

从1957年到1963年，在最后找到经济上可以接受的解决办法之前，设计队伍反复尝试了12种不同的建造"壳"的方法（包括抛物线结构、圆形肋骨和椭圆体）。"壳"的设计工作是最早利用电脑进行构造分析的工作之一。在1961年中期，设计队伍找到了一个解决办法：所有的"壳"都由球体创建而来。该办法可以使用一个共同的模具浇注出不同长度的圆拱，然后将若干有着相似长度的圆拱段放在一起形成一个球形的剖面。

"壳"由Hornibrook Group Pty Ltd建造，它负责建造了第三阶段。Hornibrook在工厂中制成了2400件预制肋骨架和4000件屋顶面板，加快了工程的进度。这个解决方案的成就在于利用预制混凝土构件从而避免了建造昂贵的模具（这同样允许屋顶面板在地上按大片的预先建造组合好，而不是在高处一个一个地拼接）。Ove Arup和合作方的工地工程师惊讶于这些"壳"在完工前使用了创新的调节型弯曲钢架来支撑不同的屋顶。在1962年4月6日，据估计悉尼歌剧院将于1964年8月到1965年3月之间完成。

阶段三：内部的设计和装潢（1967年~1973年）

一直到1966年，悉尼歌剧院建造计划的花费只有2290万元，仅是最终花费的四分之一。然而在第三阶段，设计上将会有很大的支出。约翰·伍重辞职的时候，第二阶段的工程正接近完工。Peter Hall在他辞职后取代了他的位置，Peter Hall对内部的设计和装潢担负重要责任。其他一些人也在同年接受任命，取代约翰·伍重的位置。

在建造过程中，澳大利亚政府改组，使得这位建筑师被迫于1966年离开澳大利亚，从此再未踏上澳大利亚土地，连自己的经典之作都无法亲眼目睹。之后的工作由澳大利亚建筑师群合力完成，包括Peter Hall、Lionel Todd与David Littlemore等三位，悉尼歌剧院最后在1973年10月20日正式开幕。

2003年4月，悉尼歌剧院设计大师约翰·伍重先生获2003普利兹克建筑学奖。普利兹克奖是对约翰·伍重和他的杰作的最终承认。2008年11月29日，约翰·伍重在丹麦去世，享年90岁。然而令人遗憾的是，这位悉尼歌剧院的设计大师，在他生前直至去世都没能够亲眼看到他自己的杰作。

Unit 5

Why The Towers Fell?

Teaching Guidance for Watching, Listening & Reading

Watch videos, pay attention to the *Words and Expressions* and related *sentences* and *paragraphs*.

By now you've seen the images and heard the stories about September 11 attacks.

But what really caused the Twin Towers to *collapse*?

Was their failure inevitable? Or could they have stood longer, giving occupants and emergency crews a better chance for escape? What's called a *"Progressive Collapse"*?

This *unthinkable tragedy* has come to define our times. The question now is "**Can we learn from it?**"

5.1 The Quest for 911 Attacks

NARRATOR: *By now you've seen the images and heard the stories, see Fig. 5 – 1.*

BILL FORNEY (World Trade Center Survivor): I remember a coworker saying, "Don't! Don't! Don't open the door. Don't go out there. It's fire out there. You're going to, you're going to burn up."

NARRATOR: *But what really caused the Twin Towers to collapse? Was their failure inevitable? Or could they have stood longer, giving occupants and emergency crews a better chance for escape?*

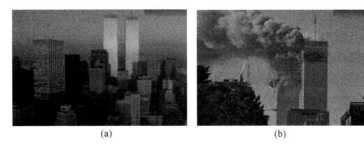

Fig. 5 – 1   Skyscraper Twin Towers

FIREFIGHTERS: When we hit the fifth floor, that's when everything happened. It was *rattling*. It was *rolling*. It was *roaring*. The floor was shaking. I remember getting knocked down the *stairwell*, landing like a *rag doll*. That's when the building started coming down.

NARRATOR: *When a blue ribbon team of forensic engineers was asked by the government to determine exactly what triggered the Towers' collapse, NOVA was there from the beginning, following their quest for answers.*

W. GENE CORLEY (Structural Engineer): What we looked for is pieces that were in the areas where fires occurred. You can get a better idea of what the strength was before the collapse occurred.

NARRATOR: *From their detailed examination of the Towers' innovative design to the search for* *forensic evidence in the molecules of collapsed steel, the investigation team has studied every possible scenario. Could one tower have collapsed for different reasons than the other? Was there something about the Towers—built to maximize rental space—that traded safety for economy?*

CHARLES THORNTON (*Structural Engineer*): A lot of people are saying that the structural engineering of the World Trade Center was miraculously wonderful, that the buildings stood up in the case of two 767s flying into it. I would tend to think they were not as successful as they could have been.

NARRATOR: *Was the damage from the explosions and massive fires too great for any building to sustain?*

MATTHYS LEVY (Author, *Why Buildings Fall Down*): As the steel began to soften and melt, the *interior core columns* began to give. Then you had this *sequential failure* that took place where it all pancaked—one after the other.

NARRATOR: *Why could only eighteen people from the impact areas or above get out alive? Was there a problem with the emergency stairs? The escape route? And perhaps most importantly, what does this disaster tell us about the safety of all tall buildings?*

JAKE PAULS (Building Safety Analyst): *Public perception* about evacuation of large buildings is that if they decide to evacuate, they will get out quickly. The reality is really something quite different.

NARRATOR: *This unthinkable tragedy has come to define our times. The question now is, " Can we learn from it?"*

LESLIE ROBERTSON (Engineer, World Trade Center): I cannot escape the people who died there. It's still, to me, up there in the air, burning. And I cannot make that go away. Why the Towers fell?

5.2 The Structural and Fire-resistant Design of WTC (World Trade Center)

NARRATOR: *Even today Ground Zero has the capacity to shock, because what happened here on September 11th still seems beyond comprehension.*

BRIAN CLARK (World Trade Center Survivor): He said, "You know, I think those buildings could go over." And I said, "There's no way. Those are *steel structures*." And I didn't finish the sentence.

NARRATOR: *People come as pilgrims to the site, to honor those who were lost and to try and understand how two of the world's tallest skyscrapers could have been destroyed so quickly. A disaster on this scale raises two crucial questions: Was the collapse of the buildings inevitable? And need so many people have died?*

MIKE MELDRUM (Ladder 6 Fire Crew): I still find it hard to believe that these buildings are missing. I can't explain what happened. I can't explain how we walked out of that building.

NARRATOR: *One place to begin the search for answers is among piles of charred and twisted steel* now lying in a scrap yard in New Jersey. Gene Corley is leading a team from the American Society of Civil Engineers investigating the precise causes of the collapse. Corley led the investigation of the Oklahoma City bombing disaster, but the magnitude and relevance of this investigation is daunting.

GENE CORLEY: I have looked at now two major terrorist attacks, and I never want to look at another one in the future. I want the findings that we have obtained from these studies to be used to develop buildings that will provide more safety for those who are in those buildings.

NARRATOR: *In the months since the collapse, the team has analyzed countless fragments of steel and pored over hundreds of hours of video tape trying to determine the timetable of the collapse and exactly which parts of the buildings failed. But to really understand how these structures performed, the team had to look back at decisions made 35 years ago, when the Twin Towers were designed and built. It all began in 1966, with a radical dream. The World Trade Center Towers were designed to be more modern, more economical and taller than any other skyscraper in the world. The lead structural engineer on the project was Leslie Robertson, then just 34 years old.*

LESLIE ROBERTSON: It was really a young person's project. It took a huge amount of energy. I did a lot of things that I don't think an older engineer would have bothered to do, because he would have had confidence in the work that he'd done in the past. And I was charging down a different highway.

NARRATOR: *Earlier skyscrapers, like the Empire State Building, used a dense grid of steel girders to support the height and mass of the structure, but they all had the same drawback.*

LESLIE ROBERTSON: The buildings of the past had columns spaced roughly 30 feet on center in all directions. And the issue with that is it worked well but it has columns in space that you would like to rent.

NARRATOR: *To increase the rentable floor space, Robertson repositioned most of the inner columns to the exterior wall, as shown in Fig. 5-2 This dense steel palisade would support half the downward weight of the building But its main task was to resist the biggest load on any skyscraper——the force of the wind.*

LESLIE ROBERTSON: That whole issue of wind engineering is the most important part of the structural design of any very tall building. Just the brute strength of it is the driving force behind all structures of tall buildings.

NARRATOR: *The World Trade Center's exterior skeleton was exceedingly strong, capable of resisting the lateral force of the wind and the unexpected force it would receive three decades later.*

BILL FORNEY: It lurched forward, back and forth. After maybe 6 to 10 movements back and forth of that building, it was over—and it was still standing.

NARRATOR: *In Robertson's design, the downward weight of the building was also supported by large steel columns around the building's inner core, which is where he placed elevator shafts, emergency stairs and other building services. But the tall vertical columns of the inner core and outer walls were like freestanding stilts until Robertson tied them together with floor trusses, see Fig. 5-3.*

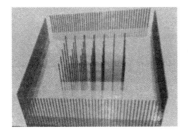
Fig. 5-2 The Structural Scheme

Fig. 5-3 Floor Trusses

LESLIE ROBERTSON: The World Trade Center is a very large project. In essence, it still boils down to a series of small pieces, and Fig. 5-4 is an example of a top part assembly of a typical floor truss.

NARRATOR: *Long and thin, these horizontal* steel assemblies *were connected by bolts to the columns at each end and then welded to the exterior columns for extra support, see Fig. 5-3. The trusses were critical for holding the buildings together, and their performance is now at the heart of the investigation into what happened. Robertson tried to save weight and costs wherever he could. He fireproofed all steel members, including the trusses, with the latest lightweight* heat-resistant foam, *see Fig. 5-5. And he kept the core area light by walling it off with* drywall *or* sheetrock *rather than concrete.*

Fig. 5-4 Steel Assemblies Connection

Fig. 5-5 Heat-resistant Foam

JONATHAN BARNETT (Professor, Fire Protection Engineering): This is very typical. We often build buildings this way, two layers of sheetrock on either side of a steel framework. It's just like you might build a wall, except we use special sheetrock that's particularly fire-resistant.

NARRATOR: *Although drywall is indeed effective at keeping fire at bay, it has one serious drawback that would reveal itself on September 11th. It's not very strong, especially when it's been heated.*

5.3 The First Unthinkable Tragedy in 1993

The designers of the Trade Center tried to anticipate every possible disaster. The Towers were the first skyscrapers ever explicitly built to survive the impact of a plane.

LESLIE ROBERTSON: We had designed the project for the impact of the largest airplane of its time, the Boeing 707, that is, to take this jet airplane running into the building, destroy a lot of structure and still have it stand up.

NARRATOR: *When the World Trade Center was opened, there was little doubt that these buildings were as safe as any skyscrapers in the world. Although they were initially criticized for their sterile, industrial look, over time they worked their way into the hearts of New Yorkers and became one of the busiest spots in the city. But their very success made them a target. In 1993, extremists attacked the buildings for the first time. The terrorists exploded a huge bomb in a parking garage beneath the Trade Center complex. The blast blew a 90-foot hole through five floors of the underground structure, killing six people and sending soot and smoke racing through the building. But the Towers stood firm.*

LESLIE ROBERTSON: The bombing, I think, created a lot of confidence in everyone's mind that the Trade Center was pretty sturdy.

NARRATOR: *The bombing did reveal at least one serious problem.*

JAKE PAULS: Remarkably, there was no study performed of the people evacuating in 1993. The *evacuation time* was something in the range of one to seven hours depending on how high you were in the building and what your disabilities were.

NARRATOR: *After the attack, exit stairs were much improved, but no one really knew if people could evacuate the towers in less time if they had to.*

5.4　The Test Came on September 11, 2001

　　The test would come on September 11, 2001. On that day British transport consultant Paul Neal was sitting at his desk on the 63rd floor of the North Tower.

PAUL NEAL (North Tower Survivor): The day was a beautiful, clear day which I'm quite sure. It was significant because it meant that the *hijackers* of the aircraft would have perfect visual conditions, so they'd been able to see those twin towers probably 60 or 70 miles away.

NARRATOR: *Down below, in the World Trade Center's underground station, the morning rush hour was underway. It's now estimated that there were just 14,000 people in the two towers at that time in the morning, far fewer than 40,000 who would normally fill them later in the day.*

DISPATCHER: Ladder 6, Ladder 6 only, Box 215, 120 Mulberry Street.

NARRATOR: *A few blocks away, the Ladder 6 fire crew was going about its normal duties when one of them heard the roar of aircraft engines.*

MATT KOMOROWSKI (Ladder 6 Fire Crew): We started pulling out of quarters and I distinctly remembered hearing from the dispatcher, "All Lower Manhattan Units respond to the World Trade Center."

SYNC: Go! Go to the Trade Center.

NARRATOR: *Inside the stricken North Tower, just ten floors beneath where the plane had hit, it was commodities trader Bill Forney.*

BILL FORNEY: There was a *high-pitched scream*. There was a tremendous change in the air pressure. The building lurched forward, back and forth. It was a scary situation. It was actually the first time I had truly ever thought that I might die.

NARRATOR: *The 767 that flew into the North Tower was larger than a 707 and was moving fast. It struck the building between the 93rd and 98th floors, instantly killing scores of people in the plane and tower. It also created a huge void across six floors on the impact wall. You can see the outline of the wing tip on the upper right. Two-thirds of the supporting columns were completely severed, but the building stood firm.*

GENE CORLEY: What happened was that the loads that were being carried by those columns arched across the opening so that the columns adjacent to the hole now started picking up the loads that had been carried by those where the airplane went.

5.5 Was the Fuel Load Considered in the Design?

NARRATOR: *Leslie Robertson's radical design seemed to have worked, but there was more devastating damage hidden inside. Although the aluminum aircraft shattered on contact with the exterior wall, the speed and force of the fragments and the intact steel engines severely damaged the columns and stairwells in the core, and jet fuel began saturating the building.*

PAUL NEAL: Almost immediately after the impact, somewhat bizarrely, I smelled an overwhelming stench of aviation fuel, Jet A1 gas, which I recognized because I'm a *private pilot* and I'm used to airfield environments. I recall smelling it and almost instantly dismissed it as being illogical and didn't have any place in the World Trade Center.

NARRATOR: *In an instant, the fuel ignited a massive fire that quickly engulfed the damaged area, and this was something even Robertson had not considered.*

LESLIE ROBERTSON: With the 707, to the best of my knowledge, the fuel load was not considered in the design. Indeed, I don't know how it could have been considered.

CHARLES THORNTON: They didn't have the *mathematical models* in the computers to model a fire as a result of the fuel in a 707. I was asked in 1986 what would happen if a plane flew into the Trade Center. And I said it would not knock the building down from the pure physics of the mass hitting the building. But we... none of us really focused on that kind of a fuel fire.

NARRATOR: *Initial reports described the fire as "Super hot" due to the thousands of gallons of jet fuel carried by the plane. But the fire experts on the study team found those reports to be wrong.*

JONATHAN BARNETT: The role of the jet fuel... although it was hot, it only lasted a short period of time. It's very similar to using lighter fluid on a *charcoal fire*. It ignites the charcoal and then burns out. Its main role was to ignite other combustibles and really start the whole space burning at once.

5.6 Whatever You Have, You Have to Try!

NARRATOR: *The fuel served to flash started the fire on several floors instantaneously. And since* sprinkler piping in the core was completely destroyed, there was no water to slow down the blaze. Even worse, when the core was struck, the building's three emergency exits were also destroyed. So 950 people above the impact became trapped with no way out of the growing inferno.

LESLIE ROBERTSON: The people above... obviously they were suffering terribly, the people who elected to take their own destiny in their hands by jumping... I mean, it must have been an incredibly awful place above the impact.

NARRATOR: *It now appeaed that people could not get past the crash area because the drywall used to protect the stairs had been blown off, leaving the staircases destroyed or in flames. One remarkable story illustrates just how weak the drywall was. The crash caused an elevator to jam between floors trapping six people in the middle of the building.*

AL SMITH (Mail Services): "If we don't get out of here, are we going to suffocate in here? Will the elevator move again?" I thought a lot of thoughts just raced through our mind at that particular moment.

NARRATOR: *They pried the doors open and found themselves facing drywall which Jan Demczur attacked with nothing more than his window cleaning squeegee.*

JAN DEMCZUR (Window Crew): Whatever you have, you have to try. And this particular time there was not brick or concrete or something, there was drywall.

AL SMITH: He took the handle of the squeegee, took the rubber out to make a device to work with. I focused on this guy digging into the wall like there was no tomorrow.

JAN DEMCZUR: I was chopping... I don't know... my hand was tight or something and I was... and the squeegee went straight through the hole. And I lost my squeegee.

NARRATOR: *Others in the elevator took over and by kicking the drywall to enlarge the hole. Al Smith, the slimmest, went through first.*

AL SMITH: I went head first, then my shoulders, which was a tight squeeze. Then I hollered back into the elevator for them to push my feet.

NARRATOR: *All six people got out alive by breaking through the weak drywall. For people higher, that weakness proved fatal. Some safety experts believe that stronger walls might have allowed many more people to get down the stairways.*

JAKE PAULS: If the stairs had been more hardened, the walls would have been less able to be breached by the collision of the aircraft. Perhaps one or two of the stairs would have survived in the impact. And that would have meant people from above could have passed through the impact area.

NARRATOR: *For 6000 people below the impact, the stairways were clear and they quickly began leaving the building. In the early stages all eyes were on the fire and the victims were stuck high in the Tower.*

CROWD SYNC: I saw it. I saw it.

NARRATOR: *No one was thinking that there might be even worse to come, but the seeds of destruction that would eventually bring the tower down had already been sown. These images reveal that spray-on fireproofing was completely blown off critical load-bearing steel, and several of the* *floor trusses were destroyed. Inside, additional trusses would have been weakened or dislodged, and fireproofing everywhere would have been* obliterated.

CHARLES THORNTON: Once the plane hit and the fragments of the plane came through the building, we knew it knocked out floors. We also knew that it knocked spray-on fireproofing off a lot of the components. Once you lose the spray-on fireproofing, you have bare steel. Once you have bare steel, you don't have a fire rating anymore.

NARRATOR: *Without fireproofing, the steel in the core was now exposed to intense heat.*

MATTHYS LEVY: So that fire caused the steel to soften up. The columns in the interior of the core began to soften, buckle and fail. And I saw that the building had really a good chance of collapsing at that point.

5.7 Firefighters' Rescue without Fear of Danger

NARRATOR: *Unaware of the danger that lay ahead, firemen began assembling to enter the crippled North Tower. Among them was the crew from Ladder 6 who faced a brutal climb of 93 floors.*

MATT KOMOROWSKI: It was very slow. We took our time going up because we have heavy gear... very crowded. The civilians were coming down on our left in a single file.

JAKE PAULS: The stairs were narrow. They were crowded and they have firemen coming up. Now, everytime a firefighter came up heavily loaded with gear, the people coming down the stairs had to stop or twist to the side, so the firefighters, if anything, tended to slow down the evacuation. In perfect hindsight, firefighters should have focused on facilitating people getting out of the building at the bottom as opposed to trying to help people at the top of the building. Those people were already lost.

NARRATOR: *Controlling the fire was also hopeless. It was too high and spread over a too big area. Had firemen even reached the* blaze, *the equipment they carried would have made very little difference. By now, more than 2000 people had managed to escape from the burning North Tower. And many occupants had also decided to leave the undamaged South Tower, a decision that probably saved countless lives. But not everyone left the South Tower. Brian Clark, a volunteer fire marshal, stayed on the 84th floor.*

BRIAN CLARK: I am strictly guessing, but I would think we were perhaps down to about 25 people left on our floor. There was an announcement coming over the system and said," If you are in the midst of evacuation, you may return to your office by using the re-entry doors on the re-entry floors and the elevators to return to your office. "

5.8 What's Called a " Progressive Collapse"?

NARRATOR: *It is not known how many, but some people turned around and went back up. Just five minutes later the terrorists struck again. A second hijacked 767 crashed into the South Tower, hitting it between floors 78 and 84. Brian Clark's office was on the 84th floor where the upper wing* *of the aircraft struck.*

BRIAN CLARK: Our room fell apart at that moment—in complete destruction. For 7 to 10 seconds, there was enormous sway in the building, and it was all one way, and I just felt in my heart that," Oh my god, we're going over. "

NARRATOR: *The plane sliced into the South Tower at an angle to the right. Unlike in the North Tower, the core was not hit dead on. But in one crucial respect the South Tower was hit in a far more damaging way than the North. It had been struck far lower down which meant the wounded section was having to bear a far heavier load, as shown in Fig. 5 -6. and Fig. 5 -7.*

Fig. 5 -6　Schematic Diagram of the North Building　　Fig. 5 -7　Schematic Diagram of the South Building

LESLIE ROBERTSON: And I kept saying to myself," What's going on inside? How bad is it inside?" And there's no way to measure it.

NARRATOR: *Although the path of the impact did not compromise the core as severely as in the North Tower, here the plane acted like a snowplow, pushing office contents and* debris *into the northeast corner and starting a raging fire at that spot. When the plane struck, there were about 2000 people left in the South Tower, 500 above the impact line and some 1500 below. For those above the crash site, two of the three staircases were completely destroyed. But a lucky few somehow found the one that was passable.*

BRIAN CLARK: So we started down that stairway and we only went three floors, and there was a group of seven of us, myself and six others. We met two people that had come up from the 80th floor—a heavy-set woman and, by comparison, a rather frail male. She said," Stop, stop. You've got to go up. "And she labored up to join us, moving very slowly. She was such a big woman. She said," You've got to go, you've got to go up, you can't go down. There's too much smoke and flame below. "

NARRATOR: *Clark then heard cries for help coming from a damaged office nearby. It was banker Stanley Praimnath. Clark pulled him free, and together they decided to chance the smoky stairs. But their progress was hampered by one of the things that was meant to protect them, the fire - resistant drywall.*

BRIAN CLARK: Drywall had been blown off the wall and was lying propped up against the *railings* here, and we had to move it, *shovel* it aside. You could see through the wall and the cracks and see flames just licking up, not a roaring inferno, just quiet flames licking up and smoke sort of eking through the wall.

NARRATOR: *Clark and Praimnath were two of only eighteen people to escape the towers from the impact zones or above. Less than a quarter of an hour after it had been hit, all the conditions for the collapse of the South Tower were in place. The huge weight of the top third of the building was bearing down on the weakening structure. Analysis of the steel from this part of the building reveals that the fire here reached 2000 degrees Fahrenheit, a temperature that would definitely have caused the steel to buckle. Inside, the fire was weakening the floor trusses. Some were starting to soften and sag, pulling on their bolted connections to the columns.*

CHARLES THORNTON: They had two 5/8 – inch bolts at one end of the truss and two 3/4 – inch bolts at the other end, which is perfectly fine to take vertical load and perfectly fine to take *shear loads*, but once the floor elements started to sag during a fire... okay... they started exerting *tension forces* because it becomes a catenary, like a clothesline, and those two little bolts just couldn't handle it.

NARRATOR: *The trusses were essential for holding the building together. If too many failed, the building would collapse.*

LESLIE ROBERTSON: I think the structures were stalwart, but they were not that stalwart. There was no fire suppression system that could even begin to deal with that event. Nothing. Nothing. So I didn't know whether they would fall or not fall.

NARRATOR: *The South Tower had now been burning for 50 minutes. The North Tower for over an hour. Office workers from both buildings were reaching at the ground level in a steady stream.*

PAUL NEAL: There was burning debris. Over this whole plaza level were, well, bodies and body parts, and I'm assuming these were the people who had been jumping.

NARRATOR: *Paul Neal made it out to the street. Others were led below ground.*

BILL FORNEY: They were ushering us forward, "Let's go. Let's keep moving." We walked along these escalators down to the tunnel system, the concourse underneath the World Trade Centers.

NARRATOR: *Seven minutes later, Brian Clark and Stanley Praimnath made it out of the South Tower and were four blocks away when they both looked back.*

BRIAN CLARK: And Stanley said to me,"You know, I think those buildings could go over." And I said,"There is no way." I said,"Those are steel structures. That's furniture and paper and carpeting and *draperies* and things like that are burning."

NEWSCASTER: We're standing next to the World Trade Towers. The police and firemen...

BILL FORNEY: A tidal wave of destruction just flowed. I remembered tightening my eyes as tight as can be, grimacing and hoping that I wasn't going to die.

GENE CORLEY: The South Tower over here and the damage to the side...

NARRATOR: *As they searched the visual record for the precise moment and trigger for the South Tower collapse, engineers Gene Corley and Bill Baker found crucial evidence in video shot by a nearby firm of architects. It reveals that much of the central core remained momentarily intact when the outer walls fell. If the core remained standing, something else must have triggered the collapse.*

CORLEY AND BAKER: It started to collapse. It spread over to here and the top of the building now was on its way down.

NARRATOR: *These pictures show that the South Tower fell away from the impact wall and toward the side where the fire had concentrated. To the team, this suggested a particular mechanism for the collapse, which the video helped confirm. The plane slammed along the eastern wall, starting a fierce fire in the northeast corner and severely damaging many of the steel columns in this area. The heat of the fire would have softened both the floor trusses and the outer columns they were attached to. When the steel became weak, the trusses would have collapsed. And without the trusses to keep them rigidly in place, the columns would have bent outward and then failed.*

CHARLES THORNTON: As you started to lose the *lateral support* due to the floors, the exterior just crumpled like a piece of paper. Or if you took a sheet of cardboard and you put some weight on it. When you take out the lateral supports, it would just bow right out.

Fig. 5 – 8   Progressive Collapse

NARRATOR: *This footage shows the process in action. A line of columns in the outer skeleton snapped. The top of the building then lurched outwards and fell. As it did so, it dislodged many more floor trusses. Once the trusses failed, the floors they were holding cascaded down with a force too great to be withstood. The result is called a "progressive collapse", as each floor pancaked down onto the one below, as shown in Fig. 5 – 8. In all, 600 people died in the South Tower, and those numbers could have been much worse.*

JAKE PAULS: Had the evacuation occurred an hour or so later when the buildings were more occupied, the story would have been very, very different. Because then the evacuation time for those extra people would have exceeded the time during which the towers were actually standing after the impact, and so there would have been many people who would be trapped in the stairways or even on the floors at the time of the collapse.

NARRATOR: It's almost impossible to overstate the shock of that first collapse.

PAUL NEAL: So I came back out onto the surface and came out into what would be my idea of what a nuclear winter would have been like.

BRIAN CLARK: The building I had worked in for 27 years was gone. And it was just a staggering thought. I mean there was then silence. People just couldn't believe it.

NARRATOR: *There was now one terrible implication. If the South Tower had fallen, the North was likely to follow. An urgent message was radioed to all firemen in the building. The Ladder 6 team had reached the 27th floor when they got the word to evacuate immediately.*

MIKE MELDRUM: We heard someone yelling on the radio, "It's time to start back down now."

MATT KOMOROWSKI: We are trained to go and save people and go into dangerous situations. And then when we were told to abandon our assignment, it was a very odd thing.

NARRATOR: *By now, most of the people who could have gotten out of the burning tower had gotten out. The men of Ladder 6 had only made it down to the fourth floor when, at 10:28, the tower came down.*

MATT KOMOROWSKI: I felt an incredible rush of air at my back.

SAL D'AGOSTINO (Ladder 6 Fire Crew): I remembered hearing the boom, the boom. As the floors were pancaking, I was hearing that.

MIKE MELDRUM: It was like standing in between two heavy freight trains in a tunnel going by you.

NARRATOR: *Over 1400 people died in the North Tower but somehow the Ladder 6 team survived.*

MIKE MELDRUM: I said,"Captain," I said,"there is a light above us." I thought it was somebody with a flash light. And I said,"What is it?" And he said,"Mick, there is a beautiful blue sky above us." And I said, "Captain, there is a 105-storey building above us." He said, "No." He said," I think we are the top of the World Trade Center right now."

SAL D'AGOSTINO: Yeah.

NARRATOR: *Nearly 3000 people perished in the attack. Four hundred seventy nine were from the emergency services. One hundred fifty seven were on board the two jets. The majority of the casualties were office workers who had been trapped in the crumbling towers.*

BILL BAKER: The fire may be related to the initial impact.

NARRATOR: *The team now believes the North Tower collapsed in a different manner than the South. The main clue lies in what happened to the TV antenna, which rested directly on top of the core.*

GENE CORLEY: Looking at the films of the North Tower, it appears that the antenna starts down just a little bit before the exterior of the building. That suggests the core went first.

MATTHYS LEVY: It was very much like a controlled *demolition* when you look at it, because the building essentially fell almost vertically down, as if someone had deliberately set a blast to take place to cause the building to fall vertically downward.

NARRATOR: *The reason the core failed first in the North Tower can be explained by the way it was hit. The 767 had smashed through the outer wall and impacted the inner core directly, damaging or destroying essential load-bearing columns and their fire protection. In this scenario, the fire would have softened the already weakened core columns to the point where they could no longer carry weight from above. When these columns finally failed, they immediately precipitated another progressive collapse.*

5.9 What was Specific Steel Components' Failure?

Knowing how the towers collapsed does not fully explain what specific components failed. For that, the investigators needed to examine the remains of the buildings.

ENGINEER: The first numbers identify which building it's in.

NARRATOR: *Most* steel components *had locator numbers stamped on the surface.*

GENE CORLEY: This part of the column extended from Floor 39 to 41.

NARRATOR: *If those numbers survive, investigators can actually pinpoint where each piece of steel came from, even in this surreal and mangled pile. Central to the investigation is finding the floor trusses and their connections.*

ENGINEER: This is the bottom piece of the truss connection.

NARRATOR: *These lightweight but critical supports have been prime suspects in the collapse, and many observers have been outspoken on this issue from the beginning.*

CHARLES THORNTON: Had the floor system been a more robust floor system with much stronger connections between the exterior and the inside, I think the buildings probably would have lasted longer. Would they ultimately have collapsed? Maybe not.

NARRATOR: *From the evidence found at the steel yards, and from computer modeling of applied forces, the team now believes the truss connections probably did fail from the force of the impact, the heat of the fire, or both. But the study concludes that there was a more fundamental reason for the overall collapse.*

GENE CORLEY: We found that the types of fireproofing that were used were damaged by the aircraft hitting them. If the fire resistance of the building was increased so that the material there could burn out before a collapse occurred, then you could come back in quickly afterwards, stabilize the building and save it from collapse.

NARRATOR: *The team concluded that the fire-resistant foam was blown off with ease. If it had remained intact, the steel would have kept its resistance to the damaging effects of the heat. And since the contents would have eventually burned out on their own, had the steel been better protected, the Twin Towers might not have fallen.*

5.10 What Does This Disaster Tell Us about the Safety of all Tall Buildings?

JONATHAN BARNETT: We have a long history of successful steel construction in this country and, in fact, the world. And one of the great successes is that under normal fire conditions we don't have building collapse. In fact, until 9/11, I was unaware of any protected steel structure that had collapsed anywhere in the world from just a fire.

GENE CORLEY: It was the combination of the impact load doing great damage to the building, followed by the fire that caused collapse. We need to look for types of fireproofing that can take the impact and can stand up to the impact and stick to the steel after the impact.

NARRATOR: *The report also points to failure of the drywall construction to protect the emergency exits in the core.*

MATTHYS LEVY: The core in concrete might have actually stood for a much longer period of time, allowing many, many more *occupants* to leave the building. It would certainly have allowed the occupants on the upper floors to have a safe passage through at least one of the vertical stairwells. The core in concrete might have actually stood through the fire and survived.

NARRATOR: *The official report suggests that, in the future, architects and engineers should consider hardening stairwells; toughening fire protection on all steel members, especially their points of connection; and creating back up supports in case key load-bearing systems fail. Given the recommendations of the study team, it is hard to imagine that these are the ruins of buildings so stalwart and strong that they actually saved people's lives. Yet this is the central conclusion of the report and its most controversial finding.*

CHARLES THORNTON: A lot of people are saying that the structural engineering in the World Trade Center was miraculously wonderful, that the building stood up that long in the case of two 767s flying into it. I would tend to think that they were not as successful as they could have been. I think the buildings, had they been a different floor design, probably would have lasted longer.

NARRATOR: *Although the World Trade Center collapse will be studied for years to come, Gene Corley stands by his team's assessment.*

GENE CORLEY: The buildings, we found, performed well. They demonstrated that they could take the hit of a large aircraft and not immediately collapse, and there was no trade off of safety for economy in construction.

NARRATOR: *In the meantime, what is left is a fierce human tragedy and thousands of people trying to come to terms with it.*

BRIAN CLARK: We lost 61 dear friends that we worked with and laughed with for years. I'm deeply saddened that they aren't here.

BILL FORNEY: You know, it scares me to think about going into a tall building. One of the big visions that I have is that the building is going to fall. And in the past, I probably would have written that vision off, thinking that could never happen. But now I know that it can happen.

MIKE MELDRUM: I ride the ferry home at night and I still find it hard to believe that these buildings are missing. I can't explain what happened. I can't explain why anybody would go to that extent. I can't explain how we walked out of that building.

NARRATOR: *But for Leslie Robertson, the man who built the World Trade Center, there is a special kind of torture: his office overlooks what was once his greatest achievement.*

LESLIE ROBERTSON: Ground Zero is a very disturbing place for me. I mean I probably have more *emotional attachment* to it than maybe any other person now alive. And I cannot escape the people who died there. Even if I'm looking down into a pile of *rubble*, it's still, to me, somehow up there in the air, burning. And I cannot make that go away.

Words and Expressions

collapse	倒塌，塌陷，垮掉	stairwell	楼梯井
rattling	嘎嘎作响	scenario	设想，方案，预测
roll	旋转	pilgrim	朝圣者
roar	吼叫，咆哮	skyscraper	摩天大楼

fragment 碎片，片段
drywall 干砌墙，清水墙，板墙
sheetrock 石膏纸夹板
hijacker 劫持者
staircase 楼梯
suffocate （使）窒息而死，（把…）闷死
squeegee 橡胶滚轴
squeeze 拥挤
stairway 扶梯
fireproofing 耐火装置
obliterate 摧毁，抹掉
blaze 火焰，烈火，火灾
debris 残骸,碎片,破片,残渣,废弃物
railing 围栏，栏杆
shovel 铲
drapery 装饰织物，布料
architect 建筑师
demolition 拆除
occupant 使用者，居住者
rubble 碎石，碎砖
progressive collapse 连续倒塌
unthinkable tragedy 难以想象的悲剧
rag doll 碎布制玩偶，布洋娃娃
blue ribbon team 权威团队
innovative design 创新设计
rental space 出租空间
structure engineer 结构工程师
massive fire 熊熊大火
interior core column 内部核心柱
sequential failure 连续失效
public perception 众所周知
Ground zero 原爆点

steel structure 钢结构
twisted steel 扭曲的钢筋
dense grid 密网格，密结构布置
steel girder 钢横梁
exterior wall 外墙
steel palisade 钢栅栏
driving force 驱动力
exterior skeleton 外部骨架
lateral force 横向力
lurch forward 向前倾斜
inner core 核心
elevator shaft 电梯井
freestanding stilt 自由式高跷
floor truss 地板桁架
steel assembly 钢部(组)件,钢装配单元
heat-resistant foam 耐热的泡沫
steel framework 钢框架
evacuation time 疏散时间
exit stairs 楼梯出口
the morning rush hour 早高峰
high-pitched scream 高分贝的尖叫
private pilot 私照飞行员
mathematical model 数学模型
super hot 超级热
charcoal fire 炭火
shear load 剪力
tension force 拉力
fire-resistant design 防火设计
floor truss 地板桁架
lateral support 横向支撑
steel component 钢构（组）件
emotional attachment 情感依附

Translation Examples

[1] As the steel began to soften and melt, the interior core columns began to give. Then you had this sequential failure that took place where it all pancaked—one after the other.

随着钢开始软化和融化，内部核心柱承载力逐渐丧失，能支撑荷载的核心柱子逐渐

减少，最后一个楼层的倒塌就导致了下面楼层的倒塌，这样楼层一个接一个连续倒塌。

[2] The buildings of the past had columns spaced roughly 30 feet on center in all directions. And the issue with that is it worked well but it has columns in space that you would like to rent.

过去的建筑都有柱子，大约与各个方向的中心距离 30 英尺。问题是它虽运行良好，但它很难有空间可供出租的。

[3] That whole issue of wind engineering is the most important part of the structural design of any very tall building. Just the brute strength of it is the driving force behind all structures of tall buildings.

整个风工程的问题是任何一座超高层的建筑结构设计中最重要的部分。它的巨大风荷载所产生的内力就是所有高层建筑结构背后的驱动力。

[4] It's just like you might build a wall, except we use special sheetrock that's particularly fire-resistant. Although drywall is indeed effective at keeping fire at bay, it has one serious drawback that would reveal itself on September 11th. It's not very strong, especially when it's been heated.

就像你可能会建造一堵墙，但我们使用特殊的石膏夹板，尤其是具有耐火性能的石膏板墙。虽然板墙确实能有效地防止火灾，但它有一个严重的缺点，在 9 月 11 日暴露出来。它强度不是很强，特别是当它被加热时。

[5] They had two 5/8-inch bolts at one end of the truss and two 3/4-inch bolts at the other end, which is perfectly fine to take vertical load and perfectly fine to take shear loads.

在桁架的一端有两个 5/8 英寸的螺栓，另一端有两个 3/4 英寸的螺栓，这对于承受垂直荷载和承受剪切载荷来说都是非常好的。

[6] The trusses were essential for holding the building together. If too many failed, the building would collapse.

桁架是支撑建筑物的关键，如果有太多的失效，这座建筑就会倒塌。

[7] It reveals that much of the central core remained momentarily intact when the outer walls fell. If the core remained standing, something else must have triggered the collapse.

它揭示出当外墙倒塌时，中心核大部分暂时保持完好。如果核心区仍然屹立不倒，一定是其他什么东西引发了崩盘。

[8] The official report suggests that, in the future, architects and engineers should consider hardening stairwells; toughening fire protection on all steel members, especially their points of connection; and creating back up supports in case key load-bearing systems fail.

官方报告建议，在未来，建筑师和工程师应该考虑加固楼梯井；加强对所有钢构件的防火措施，特别是它们的连接点；并在关键承载系统失效时建立备用支撑。

Activities—Discussion, Speaking & Writing

Presentation

Group: 5 to 7 members

10 minutes per group (Each member should cover your part at least one or two minutes).

Clearly deliver your points of the following questions to audiences.
NEED practice (individually and together)!!
Gesture and eye contact.
Smile is always KEY!! Cover your nervousness!!

Questions for discussion and presentation

A disaster on this scale raises several crucial questions:
1. Was the collapse of the buildings inevitable? And need so many people have died?
2. Was the damage from the explosions and massive fires too great for any building to sustain?
3. Could you please introduce the structural and fire-resistant design of WTC to us?
4. What's your opinion on a few survivors "Whatever you have, you have to try"?
5. What's called a "progressive collapse"?
6. What specific components failed?
7. What does this disaster tell us about the safety of all tall buildings?

Writing

Read **Why The Towers Fell**? Write a short essay independently which contains at least the following information.

9·11 事件的影响和启示

"9·11 事件"（September 11 attacks），又称"9·11 恐怖袭击事件"，是 2001 年 9 月 11 日发生在美国纽约世界贸易中心的一起系列恐怖袭击事件。2001 年 9 月 11 日上午（美国东部时间），两架被恐怖分子劫持的民航客机分别撞向美国纽约世界贸易中心一号楼和世界贸易中心二号楼，两座建筑在遭到攻击后相继倒塌，世界贸易中心其余 5 座建筑物也受震而坍塌损毁；9 时许，另一架被劫持的客机撞向位于美国华盛顿的美国国防部五角大楼，五角大楼局部结构损坏并坍塌。

"9·11"事件是发生在美国本土的最为严重的恐怖攻击行动，遇难者总数高达 2996 人。对于此次事件的财产损失各方统计不一，联合国发表报告称此次恐怖袭击对美经济损失达 2000 亿美元，相当于当年生产总值的 2%。此次事件对全球经济所造成的损害甚至达到 1 万亿美元左右。此次事件对美国民众造成的心理影响极为深远，美国民众对经济及政治上的安全感均被严重削弱。

在美国乃至全世界都有成功的钢结构建设的悠久历史。其中一个很大的成功就是在正常的火灾条件下钢结构不会发生建筑物倒塌。事实上，直到"9·11事件"，大家还没有意识到世界上任何地方有一座受保护的钢结构会因火灾而倒塌。吉恩·科利说此次事件是撞击荷载的组合对建筑物造成了很大的破坏，接着是火灾，导致了大楼的倒塌，需要研究能够承受冲击荷载的结构形式及防火材料类型，并在冲击后防火材料依然坚固，不会从钢桁架上脱落。

调查报告还指出，防火板墙的建设是失败的，没有能够保护紧急出口的核心区域能正常疏散，若混凝土的核心区域能安全存在更长的时间，允许更多的人疏散后离开大楼，那损伤也不会这么惨重，这肯定会让上层的住户有一个安全的通道，通过至少一个垂直楼梯井，混凝土的核心区域可能实际上是能保证火中的人幸存下来的唯一途径。官方报告建议，在未来，建筑师和工程师应该考虑加强楼梯井设计；加强对所有钢构件的防火措施，特别是它们的连接点；并在关键承载系统失效时建立备用支撑。

基于研究小组的建议，很难想象这些建筑的废墟会如此坚固，它们曾实际上拯救了人们的生命。然而这个报告的中心结论是整个报告中最具争议的结论。很多人都说世贸中心的结构工程是不可思议的，当有两架767飞机飞进世贸中心时，这座建筑会站得那么久。也有人会倾向于认为它们并没有它们能做到的那么成功，如果它们是不同的楼盖设计，可能会持续更长时间。虽然世贸中心的倒塌将在未来数年内被研究，吉恩·科利仍坚持他的团队评估这些建筑表现得很好，且事实证明了可以承受大型飞机的撞击，而不是立即崩溃，而且在建筑安全方面也没有任何因经济的权衡而偷工减料。

Unit 6
Hotel Skywalk Collapse

Teaching Guidance for Watching, Listening & Reading

Watch videos, pay attention to the *Words* and *Expressions* and related *sentences* and *paragraphs*.

One sultry July evening, almost 1500 people crowded inside the hotel's *atrium* for a tea dance. Suddenly part of the building collapsed, killing 114 people. It has been the worst *structural failure* in US history. And the ASCE (*American Society of Civil Engineers*) rewrote its rules to send a clear message to *structure engineers*," You are responsible for the plans that carry you stamp".

6.1 It is One of the City's Most Spectacular Buildings

The American mid-west Missouri Kansas City, in 1978, to many outsiders, it is still the mid-Western backwater best known for cattle trading. In reality, the city was emerging from a shadow of economic *recession* as a diverse commercial center. Kansas City's authorities greenlighted a major *urban* redevelopment skim, a new 50-million-dollar luxury hotel, the Hyatt Regency. The project looked like to shake off the city's outdated cow town image.

In May 1978, builders started work on the 150 meters tall, 40-storey Hyatt. As they began to face one of the buildings, local engineer firm GCE owned by Jack Golem was still finalizing the hotel's design. This new fast track approach meant the project should be completed much more swiftly than *conventional built*. In the past 10 years, Golem has created many *brilliant architectural projects*. However, the Hyatt cooperation expected a *ground-breaking architectural statement*. Turning the vision to reality, it would be one of his firm's biggest challenges yet.

Fig. 6-1 Three Skywalks

The hotel center piece would be a impressively 44 meters wide and 15 meters high *glass ceiling atrium*. Three *suspended walkways* would spend this vast space connecting the guest rooms to a conference center and shopping complex, as in Fig. 6-1. But then after 17 months, there was amajor setback. The section of atrium's glass ceiling with the size of a tennis court, crashed to the ground. Engineer Jack Golem analyzed the problem, by which he discovered the bolt connecting *ceiling panel* with the *steel roof beams* were in wrong installs, but he concluded the accident was one-off.

On July 1st, 1980, despite the setbacks, the Kansas-City Hyatt Regency Hotel opened on schedule. It is one of the city's most *spectacular buildings*. And atrium's suspended walkways quickly became a major talking point, as shown in Fig. 6-1.

As reporter Micheal Mahoney said," Your first impression was this large open airy space and these skywalks seem suspending in air, shocking people, as in Fig. 6-2.

Fig. 6-2 The Structure of the First and Third Skywalk

6.2 The Tea Dance Competition

At 6 pm on Friday of July 17th, 1981, after 12 months since it opened, the Hyatt has become one of the city's top night clubs. This night, 1500 people crowded in the famous atrium, they took part in the hotel's most popular regular events, a 1940-style tea dance competition. As dancers practiced their moves, a TV crew arrived in the lobby. Reporter Micheal Mahoney, as a staff for the local new station KMBC for a year, he came to film an light art feature item for the evening news, but it was not his ideal of the major *assignment*.

Micheal Mahoney: I hate features, I like hard news stories and I took the job in Kansas – City as an opportunity to get into a large American market. And hopefully it's some point that walking my way into what I was more comfortable with, which was hard news story.

On the other side of lobby, engineer Walter Trueblood and his wife Shirley were enjoying the eighth tea dance in the Hyatt. They've shared passion for dance since their teenage sweetheart coating in the 1950s. As loyal customers, Walter and Shirley enjoyed special benefits.

Walter Trueblood: The tea dance has been kind enough to send two free drink tickets. And I never turn down the free martini in my life.

Local mortgage broker, Frank Freeman was the first time of the tea dance. He was here with his boyfriend Roger Greasby of five years. They've been looking forward to the dance for weeks. And Roger was in his element.

Frank Freeman: He was *extrovert*, there wasn't anybody he didn't know, he could talk to anybody. You'll take them known for life, but he just met them.

Meanwhile Walter and Shirley decided against entrancing the competition. Instead, they joined dozens of other couples crowding onto the first-floor walkway, which was bird's eye view of the down floor 5 meters below, as shown in Fig. 6 – 3.

At **7:04 pm**, The band struck a popular fox *trot* and the dance competition began. Micheal Mahoney's crew filmed the top shot from the first-floor restaurant. On the first-floor skywalk, Walter and Shirley sway to the music, booming up from the atrium. Nine meters above them, dozens more watched from the third floor's skywalk. On the lobby floor, directly underneath the skywalk, Frank Freeman and Roger Greasby admired the competitor's moves, as in Fig. 6 – 4.

Fig. 6 – 3   The Couples   Fig. 6 – 4   Frank and Roger

At 7:05 pm, in the restaurant on the first-floor *mezzanine*, reporter Micheal Mahoney's cameraman ran out of power.

Micheal Mahoney: At that point, David was setting up his shot. And I reached over into a camera bag to get some fresh batteries.

Then, Mahoney heard a noise from the cross atrium.

Micheal Mahoney: I heard this real sharp *metallic* voice like pop-pop. Anyway a sound was so weird in this environment that I immediately looked up.

Walter Trueblood: There was a loud pop, and the floor dropped 6 or 8 inches. I took Shirley's arm and said I thought we needed to step off. We took about 3 steps.

6.3 The Walkways Collapsed

Micheal Mahoney traced the popping noise to the skywalk opposite him.

Micheal Mahoney: I was directly on level with the second walkway, the second walkway began to *sag*. It took just a few seconds, but it seemed like forever at that moment, they sank down, and all of sudden it just dropped.

People looked hard, as the first and third floor glass and *concrete* skywalks collapsed into the crowded lobby.

Walter Trueblood: The upper level came down, and the whole thing was down like an elevator.

Micheal Mahoney: You looked like, oh my god.

Dozens of dancers and spectators lay dead, crushed beneath 65 tons of concrete and steel, hundreds more were buried alive. Unless rescuers reached them fast, many more would die.

Moments ago, local reporter Micheal Mahoney was capturing these carefree images of a tea dance in Kansas City. Now his video camera recorded these horrific scenes of *devastation*.

Micheal Mahoney: It was an awful scene. It was just terrible, there were people that were cut badly, there were people thaty were disabled, and there were people they had to be *amputated*. And there were people they had been trapped underneath these walkways, either *smashed* to the ground, or sandwiched in between. And it was a *gruesome* scene.

Stunned survivors made *frantic* 911 calls. There needed an ambulance in Hyatt Regency, part of the rubbish pulled down a bounce of people. "Do you know how many people are injured?" "There must be at least 100."

At 7:07 pm, two minutes after the collapse, *mortgage* broker Frank Freeman was in shock. He and his boyfriend Roger were standing beneath the walkways when they collapsed. Frank was hit on back and shoulder by falling debris, but was escaped from being crushed by the narrow floor.

Frank Freeman: When it's all settled down, I was facing the skywalk and the toes of my shoes were just barely touching the skywalk on the floor.

There was no scene of Roger. A few meters away, lying under the remains of the first and third floor walkways was property valuer Mark Williams. Just seconds ago, he was getting a drink of the bar. Now he was panned to the floor under 65 tons of wreckage.

Mark Williams: My left leg was up behind my head and appeared behind right ear. My right leg was torn out of the socket and was also lying back behind me. Both of legs were twisted behind me, up behind my head.

Now engineer Walter who was on the first-floor skywalk regained his senses. There was a crashing weight on his chest.

Walter Trueblood: The only thing I could move was my arm and I was pulling my tie off, so it gave me more air.

He had no idea where his wife Shirley was, or even if she survived.

At **7:17pm,** firefighters and emergency crew all arrived. They swiftly settled a makeshift morgue in the hotel lobby. And turned the outside hotel taxi rent into a treatment center. In this concrete *tomb*, Walter had no idea if his wife Shirley was alive. Then he heard something.

Walter Trueblood: That voice I heard was from Shirley saying Walter. And I asked how she was doing.

Shirley Trueblood: I remember Walter was calling me and wondered to know that if I was all right. And we couldn't touch but we could talk.

Now broken water pipes severed by the collapse posed new threats to survivors. Water from the hotel giant tank flew in the lobby at the rate of 1000 liters per minute. Under the *rubble*, hundreds of people who survived the collapse, now faced the prospect of drowning. Mark Williams could feel the water rising.

Mark Williams: When I was starting to breathe, the water was into my nose, I remembered thinking to myself, I was at one of the highest points in the city of the Kansas City, and I was going drunk to death.

As Mark struggled to keep his head above water, rescuers were helpless to stop the flow. Kansas City's fire chief could point the problem. The hotel front door was acting like a dam, stopping the water in the lobby from escaping. He was sending *bulldozers*. They smashed the doors down. Water started to pour out of the lobby. To Mark Williams's huge relieve, the water level started to fall.

6.4 Over 2 Hours after the Collapse

At **9:30 pm,** over 2 hours after the collapse, Walter and Shirley Trueblood finally heard rescue workers nearby, and shouted for help. The rescue workers attached steel cables to the beam hitting Shirley to the floor. And began to lift it with a *crane*, they had to go slowly. One slip might led the beam to kill her.

Shirley Trueblood: They had been talking with us, and I told them that I felt like I'm about to *faint*, was that bad? And I remembered them saying" Lady, please hanging on" . He said "I'm almost there" .

At last, rescuers pulled Shirley free. Just meters away, another team dragged her husband Walter from the rubble in painful scream.

At **10:00 pm**, Frank Freeman who was finally persuaded to leave the hotel was getting treatment into the nearby medical center, when the call came in. Rescue team found a body of a man, who fits Roger's description. They brought Frank the photo to make sure.

Frank Freeman: And he did look, he didn't look like he was beaten by anything. I mean he had no cuts, there's no *bruise*. It looks like he was just lying here sleeping. And I said what's happening, they answered he had broken the neck.

6.5 Seven and Half Hours after the Collapse

At **2:30 am**, seven and half hours after the collapse, the rescuers were only finding died bodies. What they didn't realize was under the rubble someone was still alive. But property valuer Mark Williams was buried so deep that he couldn't even hear the voice from the workman above him.

Mark Williams: I was getting mad. Why wasn't somebody lifting the stuff off me? If it was me, I'll organize a group of people to lift this off me.

Then he heard something. Rescuer was drilling into the *debris* directly above him.

Mark Williams: I heard someone beat on something above my head, which might be a skywalk whatever. And I started yelling,"Hi, I'm under here".

But the noise of jackhammer drummed out Mark's frantic screams. He survived from walkways' collapse and threaten of drowning, But now Mark Williams faced the unthinkable serious injury or even death at the hands of rescue workers.

Mark Williams: The jackhammer came through the rubble again, and went between my legs. And now I was thinking they were trying to fillet me. The next one they would come right through the middle of my back. And I should be braced for.

Then as drilling team aligned the jackhammer for another run, somebody finally heard Mark's desperate cries.

Mark Williams: I remembered a guy said,"My god, there is somebody alive under here".

6.6 Almost 10 Hours after the Collapse

At **4:30 am**, fire fighter Ray Wynn pulled Mark Williams from the pile of debris. Almost 10 hours after the collapse, he was the last person found alive. But he was horribly injured.

Ray Wynn: You can see the sole of his shoes, back here. And when he set up, and I was shocked. This guy was alive. I was so happy to see him.

At **5:00 am**, an ambulance crew rushed Mark to hospital. He was still conscious, but his back was broken, and his kidneys were failing.

Mark Williams: I do remember lying there, and my mother asked the doctor,"is he gonna live? "And the doctor said, "I don't think he's gonna make it. "I set up, and I said" like hell I'm not gonna make it. I'm going to hunt ducks on October 27 at hunter season".

As Mark Williams underwent the emergency surgery, the *enormity* of the Hyatt *tragedy* emerged. 114 people have been killed, a further 186 were injured, many of whom severely. It's the deadliest structural failure in American engineering history. Across the United States, people wanted answers. How can a key part of prestigious public project like the Hyatt Hotel simply fall apart?

Now by reviewing the disaster, and by deeply investigation, we'll reveal what really happened of the Kansas City Hyatt. How were so many people died, and why the hotel's sophisticated skywalks gave way with such devastating consequences. Advanced computer simulation will take us where no camera can go, into the heart of the disaster zone, as shown in Fig. 6 – 5.

Fig. 6 – 5 The Broken of Skywalk

6.7 Is it Due to the Spread of the Construction or the Faulty Materials?

On **Monday July 20th, 1981**, three days after the disaster, relatives held the first funerals for the victims. Kansas City was in mourning. Hyatt, the hotel operator, and the building's owner, the Crown Center Redevelopment Cooperation started their own investigation into the tragedy. But Kansas City's mayor Richard wanted independent public inquiry. He asked the National *Bureau* of Standards (NBS) to step in. The NBS is a highly regarded independent federal body, that provides scientific analysis for the US government. But its investigation could take months. Local newspaper——the Kansas City Star decided to shortly cut the process. The paper wanted a local structure engineer Wayne Lischka to become its undercover investigator.

Wayne Lischka: I was surprised by the phone call, and a little *apprehensive* about taking this job. On one side, I wanted to do, I wanted to get involved to see what really happened.

Lischka knew the Hyatt was a fast track project, completed in just under two and a half years. Did the spread of the construction cause short and cheap work? Or did the faulty materials, defective steel or *poorly mix concrete* cause the collapse? If so, many more buildings under construction in booming Kansas City maybe at risk. The answers lay in the rubble inside the Hyatt atrium. Three days after the collapse, the building's owners decided to let the press inside the ruined atrium. But the reporters among them including undercover detective Wayne Lischka, soon found there was a catch. The journalists were only allowed to view the wreckage from 30 meters away. But the Kansas City Star guessed that Lischka might be kept away from links, and hatched an *ingenious* plot. Lischka reserved a secret weapon, which was a photograph with a telephoto lens. Lischka could see the tangled wreckage on the first and third floor walkways where they fell on the lobby floor. Overhead, all the remains of original structure were six 5 – meter – long hang rods, still fixed to the ceiling. It was enough to allow Wayne Lischka to work out the walkway design. The upper walkway would've been suspended from the ceiling by these set of rods. Lischka realized that there must be a second set of rods, bottom into the upper walkway that carried lower walkway 9 meters below. More importantly, the ceiling hang rods are the crucial clues to where the failure happened.

Wayne Lischka: The first thing that caught my eyes was that the rods were still hanging from the ceiling.

Since the rods were *intact*, Lischka believed that the structure must be failed at the point where the upper walkway was connected to the hang rods. Did the faulty fixing at these critical connection points fail? To find out, he would need to exam the *wreckage* more closely, something that the builder's owners would not allow. Lischka hoped that the photographs could review more, but now he had a bring way. City Hall must hold a set of Hyatt blueprints. They would detail the fixings used, and might hold some clue to their apparent failure. But City Hall told Lischka he couldn't see the blueprints. Librarians were in the process of cataloguing them. His undercover investigation for the Kansas City Star was a brick wall. While Lischka's undercover investigation stopped, the official investigation got to gear. A team of four scientists from the National Bureau of Standards flew in to the Kansas City. Among them was the leading investigator Edward Pfrang. His first priority was to examine the Hyatt collapsed walkways. But when he arrived at the Hyatt, Pfrang and his NBS team were shocked and puzzled by what they found. The lobby was empty. The wreckage of the walkways and the crucial evidence that it contained has disappeared.

6.8　Is the Load too Large? Is it Due to the Harmony Vibration?

Scientists flowing from the National Bureau of Standards (NBS) to investigate the Hyatt Regency skywalk collapse hit a problem that the walkway wreckage has gone. The building's owners had taken it away for private analysis. For lead investigator Dr. Edward Pfrang, Chief of the Bureau's structure division, it was a massive thunderclap. "You know the debris has been moved, we have not had a chance to see that".

Edward Pfrang: We were shocked to have it removed, and very disappointed.

Pfrang immediately *petitioned* the *circuit court* of Jackson country, for access to the wreckage which lay in the warehouse downtown. If the court denied the NBS's request, its independent investigation of the Hyatt disaster was died in the water. The NBS team stared to examine eye witness' accounts for clues to the cause of the collapse. Many said that Hyatt walkways were over crowed that evening.

One witness: They encouraged people dancing in the walkways, they said to use the entire lobby as a dance floor, which everybody was doing.

Some experts said that people dancing on the walkways, may have put them on the unforeseen stress.

Pat, Hyatt Hotel's president rebutted the claim that walkways were overloaded. NBS Investigators examined Kansas City's building codes, they found that the codes required public structures to be capable of carrying a load of 488 kilograms per square meter. The third and first floor walkways should be strong enough to carry at least 1280 people. But no one knew exactly how many people at all that evening. Then the NBS investigators got a break. They learnt that the news reporter Micheal Mahoney's unique *footage* of the tea dance on the night of disaster. Investigators pulled over the footage. They discovered that several of Micheal Mahoney's shoots focused on the famous skywalks.

Edward Pfrang: It was like being given that information was a treasure, which was very, very useful.

Investigators worked out that at the time of the collapse, there were 40 people on the first floor walkway, and 23 on the third floor walkway, 63 in total. They calculated that 63 people would only exert a load of 83 kilograms per square meter, a fraction of the possible maximum load. The number of people alone was clearly not enough to overload the walkway. But could just 63 people still trigger disaster if enough of them would be dancing on the walkways?

In 1940, news camera captured the Tacoma Bridge tearing itself apart. High winds created a violent sway motion in its structure, ultimately causing the bridge to collapse, as shown in Fig. 6-6. It's an extreme example what engineers call "*harmonic vibration*". Now the NBS asked could people swaying on the Hyatt walkway have created a similar effect. Experts knew that all structures, no matter how solid it is, *vibrate* in perceptively at their own individual frequency. If the people on the walkway would move at the same frequency, it could cause harmonic vibration. That would set up a wave emotion in the walkway that could deform the structure, and untimely cause it to fail. To determine if harmonic vibration did cause the collapse, the NBS team needed to establish the *natural frequency* of the walkways and the *tracking frequency* of the people they carried, that was the *tempo* of which they were dancing or swaying to the music. Investigators learnt that the band was playing a popular fox trot at the time of the collapse. When they analyzed

(a)   (b)

Fig. 6-6 The Tacoma Bridge

the tune, they discovered it had a tracking frequency at 1.1 hertz or beats per second. They compared it to the natural frequency of the surviving identical second floor walkway. They discovered that its natural frequency was 7.1 hertz, which was more than 6 times faster. Since the two frequencies was different, harmonic vibration couldn't be affected in the disaster. It was a dead end. The NBS's investigation stopped.

Meanwhile, engineer Wayne Lischka under unofficial investigation for the Kansas City Star "got a shot in the arm". The Hyatt building plans were now back to the City Hall after *cataloguing*. He hoped that might hold some clues to the walkway's collapse. But what he found in City Hall came a bomb shock.

Wayne Lischka: I was shocked at that point. It only took a few seconds to realize that what was under plans and what was built on the field was not the same thing.

6.9 Why was the Box Beam Connection Point Failed?

The design plans filed by Jack Golem engineering firm GCE, showed the first and third floor walkways hanging from the ceiling by a series of 14 - meter - long hang rods. But Lischka knew from his visit to the Hyatt that this was not how the walkways actually were built. The rods still hanging from the ceiling were simply not long enough to reach the lower walkway. A second set of rods bolted into the upper walkway must carry its twin below. He has already *deduced* that the structure failed at the point where the upper walkway was connected to the ceiling rods. Now he was beginning to understand the reason. According to the plans, this *critical connection point* was only meant to bear the upper walkway. However, in the built structure, it carried the weight of the lower walkway as well, which was twice of intended load.

Lischka needed to know whether engineer Jack Golem changed the connection point design to take account of this extra load. Hours later, the newly developed photographs of the walkway *wreckage* gave him his answer. They clearly showed that several of the upper walkway connection points were dramatically failed. Three *cross beams* formed the walkway's basic structure, at the connection point, the hang rods passed through the end of each beam or channel and were secured by nuts underneath. It started obviously to Lischka why this connection point design was not up to the task.

Wayne Lischka: In Fig. 6 - 7, these are basically 2 eight - inch channels similar to the one

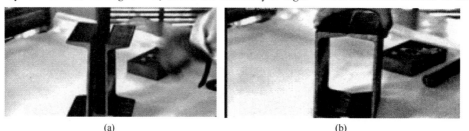

(a)   (b)

Fig. 6 - 7 Two Ways to Form a Box Beam

used in the skywalk. Normally if you wanted to make a walkway out of two channels, what they would've done was to turn them back to back, and take the rod between them. But for Hyatt, what they did was that they took two channels and turned them toe to toe. And then they welded them to form a *box beam*. Welding the two pieces together makes it very weak in the walls.

Welded box beams were adequate to provide the walkways basic *framework*. But using them to form a connection point posed a serions problem. They must bear the entire loads of both walkways. But the nature of the structure transferred the huge force to the beam's weakest and thinnest points——its walls. Lischka believed the connection point design was the recipe for disaster.

Wayne Lischka: Putting two channels toe to toe is a little absurd both at that time and at present.

Lischka published his finding in Kansas City Star. If he was right, the collapse was not caused by faulty materials, but by bad design. But without access to the walkways themselves, it remained no more an interesting theory. Then 12 days after the collapse, the circuit court of Jackson country issued a crucial judgment: It granted the NBS investigation team access to the remains of the walkways locked up in the Kansas city *warehouse*. The investigators' first priority was to confirm where the walkways gave way. Among the wreckage of the upper walkway, they had knocked all 6 of the critical connection points and made a disturbing discovery. On each of them, the welds on the box beam have folded inwards and upwards, allowing the *hang rod*s to pull right through. One box beam named 9UE showed signs of the longer term damage than the others, as in Fig. 6 – 8, suggesting that it gave way first. It looked like Lischka was right. The box beam design of the connection points was the walkways' killing injury.

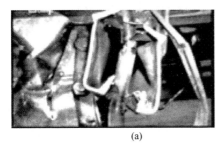
(a)
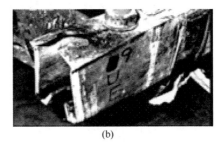
(b)

Fig. 6 – 8 Broken of Box Beam

Edward Pfrang: We are just looking at the box beams. If you were here, you'll shake your head and say I shouldn't do it that way.

6.10 The Experimental Test Showed That Connections were the Cause of the Disaster

Both Lischka and Pfrang thought the failure on the box beam connections was the cause of the disaster. But unlike Lischka, Pfrang could put a theory to the test. In the national laboratory of Gaithersburg Maryland, investigators attached *steel box beams* to the *hang rods*, to exactly replicate the connection point. They weighed every *scrap* of the walkway debris, and calculated that the total load of connection was 8150 kilograms, over 8 tons. Firstly, they wanted to find out if the *critical connection point* was able to take the weight of the walkways without people. The scientists subjected the first replica to a constant 8 tons of load. 4 seconds in, there was no change. 20 seconds in, they detected movement. After 52 seconds, it stopped. Under the weight of walkways alone, the walls of the beams started to bow, but the connection did not give way. But the investigators knew that there were 63 people on the walkways when they fell. They estimated that these would add extra tons, bringing the total load to just over 9 tons. They repeated the experiment with the second *identical* connection. This time they would gradually add the extra weight to observe exactly what happened to the connection points. At 8070 kilograms, the side walls *distorted* further and the base started to give way. At 8160 kilograms, the weld gave way with a crack. And then at 8255 kilograms, it failed completely. The base of the connection which was the point bearing the greatest stress, simply folded like cardboard. The bolts connecting the walkway to its hang rods were ripped clean through. Under its steady mechanical exerted force, the connection actually gave way under 815 kilograms shorter than the estimated load. The NBS's conclusion was clear cut and deeply shocking. The third floor walkway connections were not strong enough to carry the weight of two walkways and people. In fact, they were so weak that they had just one third of load capacity required by Kansas City building codes.

Edward Pfrang: From the day that they were built up, they were a disaster waiting to happen.

6.11 Whose Fault is the Fatal Flaw of the Skywalk?

The investigation took 10 months, it reviewed in chilling detail exactly on what caused two walkways to *tear loose from* their supports higher above crowded dance floor, leading 114 Friday night revelers stuck from disaster.

At 7:00 pm, 5 minutes to disaster, the Hyatt Friday night tea dance was filled with 1500 party goers, dozens of people flocked onto the third-floor walkway to escape the crowd. Beneath their feet, the walkway's six connection points struggled to carry the nine-ton load. At the walkway's eastern end, the connection point of the end of box beam 9UE had been gradually deforming for 16 months. Beneath the walkways, people were totally unaware of the danger overhead.

At 7:04 pm, 1 minute to disaster, Shirley and Walter Trueblood joined the crowd on the first floor walkway. Seconds later, the base of the box beam 9UE buckled, as its weld snapped, people heard the popping from out of 30 meters away. As the base of the beam folded, the hang rods ripped free. As the restraining nuts smashed into the top of the beam above with a sharp crack, the walkways dropped down sharply. 2 seconds later, disaster struck. Box beam 9UE failed. The five remaining connections on the third floor *instantaneously* broke. The third floor walkway *plumped down* at over 15 meters per second. 65 tons of concrete and steel crashed to the lobby floor, in just 1.6 seconds.

A *fatal* floor in the construction of the walkways was responsible for the death of 114 people. The job of the NBS's team was solely to explain technical reasons for the collapse. But the victims of the tragedy still wanted answers to the wider questions: How could a public structure be built without anyone realizing it was not up to the job? And who was to blame to the walkways' fatal floor? The full story would emerge a series of sensational court cases to place in the wake of the disaster. What they reviewed would shock America, and transformed the way of building's built and design in the US forever.

Official investigators have reviewed that Kansas City Hyatt skywalks collapsed because the way they were connected to the hang rods was fatally flawed. The critical connection points had just one third of the load capacity required. The spotlight now forced on Jack Golem, head of the engineering firm GCE which designed the Hyatt. In the wake of the collapse, the American Society of Civil Engineers (ASCE), and Missouri State Board investigated Golem's role. It was emerged that it was Hyatt architects who requested box beams for the walkways. Their motives were *aesthetic*, box beam was the easiest type of beam to conceal with plasterboard. But as chief engineer, it was Golem's job to insure that all aspects of the walkway was *structurally sound*. So, what went wrong? Jack Golem told investigators that one year into the Hyatt built, his firm GCE sent joints of the walkway and its connection points to the *fabricator* Haven Steel. Haven Steel said that the single hang rods of design was too hard to build, and proposed two sets of rods instead. One month later, GCE gave the new walkway design its *stamp of approval*. Investigators were mystified, Golem's concept for the connection point was fault from the start, and the new design doubled the load they would carry. But there was no stage that GCE attended to *rectify* the problem. Why wasn't the mistake spot? Now Golem made a staggering admission. Nobody in GCE ran any calculations on the strength of the walkway connections. He maintained that it was custom and practice of Missouri for fabricators to do the calculations. While Haven still insisted that it was the engineer's job to check the design. And investigators discovered that Golem had another opportunity to avoid *catastrophe*. When part of the atrium ceiling collapsed during the built, Golem's firm did check connections in the atrium. But nobody checked the walkway connections. If they had, it was almost certain the problem would have *come to light*.

6.12　ASCE Rewrote Its Rules

The ASCE suspended Golem. After 5 years of *litigation*, the Missouri State Board held GCE responsible for the collapse, and revoked Golem's license to practice in the State of Missouri. Golem believed that his firm GCE behaved correctly, but except the buck stopped with the nominal engineer. He took ultimate responsibility for the mistake.

The Kansas City Star won a Pulitzer Prize for Wayne Lischka's report on the Hyatt skywalk collapse. Lischka is still working as an engineer in Kansas City. Mark Williams spent two months in intensive care and endured many months more of painful rehabilitation. But on the opening day of the duck hunting season, he made it.

Mark Williams: I made it. My physical *therapist* carried me on his back through the mush, getting me to the duck blend on that opening day.

Despite suffering a broken back and having both legs torn out of their *sockets*, Mark made a complete recovery. Walter and Shirley Trueblood also recovered from their serious injuries, they're still king dancers.

The Hyatt skywalk disaster transformed the engineering safety in the United States. In the wake of the tragedy, Kansas City completed overhauled its building regulations, requiring that all load bearing calculations be checked by a city appointed engineer. And the American Society of Civil Engineers rewrote its rules to send a clear message to structual engineers," **You are responsible for the plans that carry your stamp**". Today the Kansas City Hyatt has been transformed, which no longer has suspended walkways. In replace, there stands a single span supported by solid columns, allowing people to have a new sense of safety and security who pass through this historical site.

Words and Expressions

atrium	中庭，天井	tomb	坟墓死亡
recession	经济衰退，经济萎缩	rubble	碎石，碎砖
urban	城市的，都市的，城镇的	bulldozer	推土机
conventional	传统的，习惯的	crane	起重机，吊车
suspended	悬，挂，吊	faint	昏倒
panel	板，镶板	bruise	使挫伤，撞伤
beam	梁	debris	残骸，碎片，破片
spectacular	壮观的，壮丽的，令人惊叹的	enormity	巨大，深远影响，严重性
assignment	（分配的）工作，任务	tragedy	悲惨的事，不幸，灾难
extrovert	性格外向者，活泼自信的人	bureau	办事处，办公室，机构
trot	慢跑，小跑	apprehensive	忧虑的，担心的，疑虑的
mezzanine	中层，夹层	ingenious	巧妙的，新颖独特的，巧妙的
metallic	金属般的，金属制的	intact	完好无损的，完整
sag	中间下垂，下凹，弯曲	wreckage	残骸，废墟
concrete	混凝土	petition	请愿
devastation	毁灭，破坏，蹂躏	footage	连续镜头
amputate	切断，截肢	vibrate	使振动，使颤动
smash	猛烈撞击，猛烈碰撞	tempo	拍子
gruesome	令人厌恶的，恐怖的	catalogue	编目录
stunned	受惊的，昏迷的	deduce	推论，推断
frantic	紧张忙乱的，手忙脚乱的	critical	极重要的，关键的
mortgage	抵押贷款	inch	英寸

weld	焊接，熔接，断接	ground-breaking architectural statement	突破性建筑成就
framework	构架，框架，结构	glass ceiling atrium	玻璃天花板中庭
warehouse	仓库	suspended walkways	悬空走道
scrap	废料，废品	ceiling panel	顶板
identical	完全相同的，同样的	steel roof beam	钢屋架梁
distort	使变形，扭曲	spectacular building	壮观的建筑
instantaneously	即刻	poorly mix concrete	低掺量混凝土
fatal	致命的，灾难性的，毁灭性的	circuit court	巡回法院
aesthetic	审美的，有审美观点的	harmonic vibration	和谐振动
fabricator	制造者	natural frequency	自然频率，固有频率
rectify	矫正，纠正，改正	trading frequency	步行频率
catastrophe	灾难，灾祸，大灾难	cross beam	横梁
litigation	诉讼，打官司	steel box beam	钢箱梁
therapist	治疗专家	hang rod	吊杆
socket	槽	critical connection point	临界连接点
structural failure	结构性破坏	tear loose from	从…撕下
American Society of Civil Engineers	美国土木工程师协会	plump down	突然坠落
structure engineers	结构工程师	structurally sound	结构坚固
conventional built	常规建造	stamp of approval	盖章核准
brilliant architectural project	辉煌的建筑工程实例	come to light	显露

Translation Examples

[1] The hotel center piece would be a impressively 44 meters wide and 15 meters high glass ceiling atrium. Three suspended walkways would spend this vast space connecting the guest rooms to conference center and shopping complex.

饭店的中央部分，将是引人注目的44米宽、15米高的玻璃屋顶中庭，三条悬空的甬道将横跨这块偌大的空间，可以从客房通往会议中心及购物建筑群。

[2] Now broken water pipes by the collapse brought new threats to survivors. Water from the hotel giant tube flew in the lobby at the rate of 1000 liters per minute. Under the rubble, hundreds of people who survived the collapse, now faced the prospect of drowning.

因倒塌而遭切断的破裂水管给幸存者带来新的威胁。从饭店巨大水槽流出来的水，以每分钟1000多公升的速度淹没大厅。天桥倒塌后，埋在瓦砾下的几百个生还者现在可能会被水淹死。

[3] Lischka believed that the structure must be failed at the point where the upper walkway was connected with the hanging rods. Did the faulty fixing at these critical connection points fail?

利西卡认为在上部人行道与吊杆相连的地方结构肯定发生了故障，是在关键连接点上（不合要求）的连接件出了问题吗？

[4] Experts know that all the structures, no matter how solid it is, vibrate in perceptively at its own individual frequency. If the people on the walkway would move at the same frequency, it could cause harmonic vibration which would set up a wave emotion in the walkway that could deform the structure, and untimely cause its failure.

专家们知道，所有的结构，无论如何坚固，都以自己的频率感知振动。如果人行道上的人以相同的频率运动，就会产生和谐振动。这会在人行道上产生波动，使结构变形，并且过早地失效。

[5] Now he began to understand the reason. According to the plans, this critical connection point was only mentioned to bear the upper walkway. In the built structure, it carried the lower walkway as well, which was twice of intended load.

现在他开始明白原因何在，根据规划图，这个关键的连接点，只是要承载4楼甬道的重量，但盖出的结构物，却要同时承载2楼甬道的重量，比预计的荷载超出1倍。

[6] Official investigators have reviewed that Kansas City Hyatt skywalks collapsed because the way they were connected to the hanging rods was fatally flawed. The critical connection points had just one third of the load capacity required.

官方调查员已经查明，堪萨斯城凯悦饭店的天桥倒塌是因为连接吊杆的方式出现致命瑕疵。关键连接点的承重能力只达到规定的三分之一。

[7] When part of the atrium ceiling collapsed during the construction, Golem's firm did check connections in the atrium. But nobody checked the walkway connections. It they did, it was almost certain that the problem would come to light.

施工期间，中庭有部分天花板坍塌时，格兰的公司检查过中庭的连接点，但没有人检查甬道的连接点。要是检查的话，就一定会发现这个问题。

Activities—Discussion, Speaking & Writing

Presentation

Group: 5 to 7 members

10 minutes per group (Each member should cover your part at least one or two minutes).

Clearly deliver your points of the following questions to audiences.

NEED practice (individually and together)!!

Gesture and eye contact.

Smile is always KEY!! Cover your nervousness!!

Questions for discussion and presentation

A disaster on this scale raises several crucial questions:

1. Did the spread of the construction cause short and cheap work? Or the faulty materials, defective steel or poorly mix concrete caused the collapse?

2. Did the faulty fixing at these critical connection points fail?

3. Could just 63 people still trigger disaster if enough of them would have dancing on the walkways?

4. How could a public structure be built without anyone realizing it was not up to the job?

5. Who was to blame to the walkways' fatal floor? Why wasn't the mistake spot?

Writing in Groups

Read **Hotel Skywalk Collapse.** Divide the following text into four parts. Write a short essay in English (in groups) which should be based on the following information.

堪萨斯城凯悦酒店坍塌事故

1 灾难的发生

1981年7月17日的堪萨斯城凯悦酒店天桥坍塌事故共造成114人死亡、216人受伤，是当时全美死亡人数最多的工程事故，直至被2001年"9·11"事件所超越。其影响之深，在27年后的7月27日，《堪萨斯日报》仍以"for many, a memorial long over due"为标题悼念该事件中的受害者。而由《堪萨斯城星报》主导的对事故原因的调查更获得了美国新闻界最高荣誉奖项——普利策新闻奖。

犹如蜜蜂选择复杂的六角形结构建造蜂巢，鸟类会被颜色鲜艳的异性吸引，在经济发达的社会，人们更喜欢挑战更大、更高、更复杂的东西。与昆虫相比，所有这些超出工程学范围的考虑因素也许会使工程师的任务更令人兴奋，但同时也一定没有那么多的经验可供借鉴。

开业于1980年7月1日的堪萨斯城凯悦酒店由三部分组成：1个40层高的塔楼部分，一个功能区，还有一个中庭。中庭是一个大型的开放空间，长44m，宽36m，高15m。3条悬空的人行天桥分别位于中庭的二、三、四层，用于连接功能区和塔楼部分。这一精巧的布局使得中庭享有开阔的空间，而人们又能够自如地穿梭于酒店各区域之中，如图1所示。

二层人行天桥和四层天桥通过吊杆相连，三层走道在另一边与之相望。7月17日当天中庭聚集了大多数来客欢快地跳舞庆祝，无缘参入其中的客人仍可以在人行天桥上驻足观看，分享这热闹的气氛。当地时间7点零5分，大约1500名民众正聚集在一层中庭兴高采烈地享受一场晚会，另有约20人在二楼，21人在三楼观看。突然，连接四楼天桥的钢质吊杆爆裂，整个四楼走道失去支撑，连同在上面观看舞会的16人一同坠落至二层走道，并与二楼走道一起跌到一楼中庭。整个坠落过程不过几秒，却导致百人罹难，以及更多的人被埋在重达60吨的碎玻璃、废铁和混凝土中。更严重的是，人行天桥的坍塌导致酒店水管的损毁，大量的水涌入一楼大厅，进一步威胁着幸存者。

搜救行动持续了14个小时，多支团队参与救援行动，多种工具和医疗设备被想方设法运进酒店。Waeckerle教授作为搜救负责人，在短时间内做出决定：轻伤者送出酒店，部分重伤者在酒店内得到治疗，受伤最严重者则只得到了止疼用的吗啡——因为人手紧迫、药物有限、时间紧急，对他们的拯救被排在了最末。主要力量用于排除接下来会发

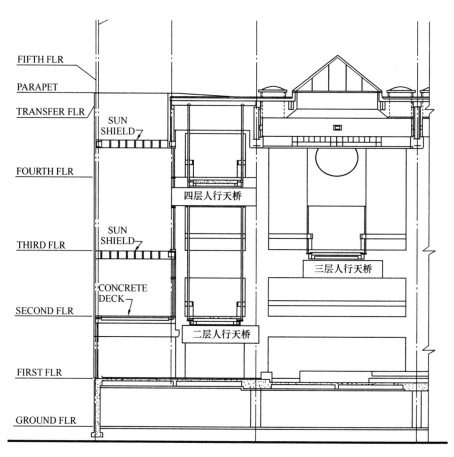

图 1　中庭示意图

生的坍塌和解围被困的人们。最终，29 人幸运地获救。

2　事故调查

凯悦酒店集团作为世界顶级的跨国酒店集团，素以豪华、舒适及人性化服务驰名。这座位于堪萨斯城的酒店在发生事故时不过刚刚运营了一年，却出现了如此严重的事故，公众强烈要求对事故展开调查。包括美国国家标准局（National Bureau of Standards，是一家属于美国商务部的非监管机构）在内的多个团队对此次事故开展了调查。其中，身为职业建筑师的 Lischka 先生接受了《堪萨斯城星报》的邀请，以记者的名义进入事故现场进行隐秘地勘察。

Lischka 先生在事故现场看到连接四楼天桥的吊杆完好无损，因此四楼吊杆的失效可以暂时被排除。而仔细排查现场保留的其他吊杆，一个惊人的事实浮出水面。

美国工程设计公司 GCE 公司，是凯悦酒店项目的设计团队，他们负责完成钢结构的设计和图纸。GCE 公司设计的人行天桥是这样的，沿整个天桥长边方向，两组工字钢（W16×26）分别支撑在天桥的混凝土走道底板下方；沿天桥的短边方向，箱梁则横向支撑在混凝土走道底板的下方，并用于固定吊杆。箱梁由两根 MC8×8.5 的 C 型钢焊接而成。如图 2 所示。

GCE 的原设计中，第二层和第四层走道是被一根吊杆连接的，这根吊杆固定在房顶

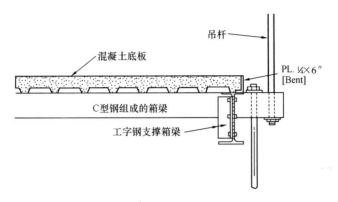

图 2　人行天桥结构示意图

上。而现场找到的吊杆长度则远小于连接两层走道所需要的长度。Lischka 先生于是重新审核了该项目的施工图纸。他发现,在施工图纸上,第二层、第四层天桥变成了由**两根吊杆共同连接**。两个螺母一上一下分别固定连接四层天桥的两根吊杆,向下的吊杆连接着二层天桥,向上的吊杆仍连接着屋顶。如图 3 所示。

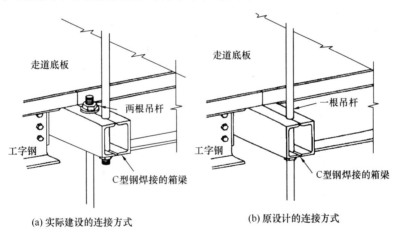

图 3　走道连接处的设计变更

原来,为了便于施工,吊杆的制造商和施工方,美国 Havens Steel Company Professional Fabricator(以下简称 Havens 公司)要求变更设计为用两根吊杆分别连接四楼和二楼。这样施工方可以免除把一根长吊杆穿过两层走道的麻烦。于是设计被改变了。**最终的施工图纸也被改变了。**

在 GCE 的原始设计中,也就是一根吊杆贯穿二层和四层走道的设计中,假设每一层走道的自重和该走道上行人的重量之和为 P,那么四层螺栓处的受力为 P,即四层螺栓处只承受该层的重量。同理,二层连接处螺栓也只承受二层走道的自重和行人重量 P。而更改的设计中,新增了一根吊杆连接二层走道,也因此,这根吊杆把二层走道及其行人的重量 P 传递到了四层走道上,四层走道的螺栓处承受的重量变成了四层走道及行人的重量 P 与二层走道及行人的重量 P 之和,也就是说变成了 $2P$,是原设计中承受重量的**两倍**。如图 4 所示。

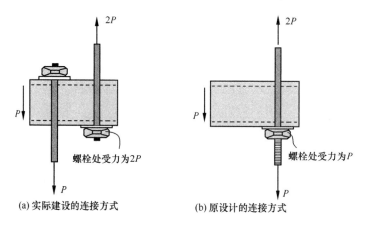

图 4　受力示意图

1979 年 2 月 16 日，GCE 收到了 42 张施工图，包括变更之后连接处的图纸。1979 年 2 月 16 日这些图纸被盖章确认并交给现场用于施工。

1980 年 7 月 1 日酒店正式开业。1981 年 7 月 17 日，天桥坍塌事故发生，四楼天桥最中间的箱梁因为其靠近东面一端连接处上的螺栓失效而迅速下滑坍塌，该处箱梁的损坏使得原本该处承担的重量转移到相邻各处连接处，其他几处连接处也纷纷失效，最终，失去了支撑的四楼天桥笔直地坠落，砸在正下方的二楼天桥上，巨大的冲击力远远超过了二楼天桥能够承担的重量，二楼连接处也断裂失效，整个四楼天桥、二楼天桥连同它们上面的人一起坠落至一楼中庭。悲剧就是这样发生的。

现场观察到的箱梁的严重扭曲变形说明了这一切破坏都是从四楼天桥箱梁上的连接处开始。据美国国家标准局的估算，事故发生时，四楼天桥连接处承受的荷载为 93kN，而美国国家标准局对天桥进行复制后进行的实验表明，该连接处最多只能承受 83kN 的力，这一数值，不仅小于事故发生时实际施加在该连接处的荷载（93kN），也远远达不到美国堪萨斯城当地建筑规范对该连接处的设计要求。根据美国堪萨斯城当地的建筑规范，该连接处必须承担 302kN 的荷载而不出现任何结构问题。即使是 GCE 的原始设计，这个数值也无法被满足。这个设计缺陷，出现在了所有的三条人行天桥上。诚如国家标准局的调查报告所说，人行天桥只能够承受自身的重量和一点点其他荷载。人行天桥从被建造出来的那一天起就等待着坍塌的危险到来。

3　责任判定

基于本次事故，由美国土木工程师协会 ACSE 出版的用于指导调查结构事故的工程师手册中对连接处失效做出了如下定义：

"由连接处失效导致的整体结构的坍塌发生在完全没有或者仅有少量的额外承载强度的结构上。当出现多个强度很低的连接处时，其中一个连接处的失效会导致相邻的连接处纷纷失效，最终致使结构坍塌。"而连接处失效的主要原因包括：① 对于施加的荷载计算失误导致的不足够（insufficient）的设计。② 由于结构截面发生突然变化导致的应力集中。③ 对于旋转和位移计算的错误。④ 连接处的材料性能退化。⑤ 没有考虑到材料在生产过程中产生的大量残余应力。

显然，凯悦酒店坍塌事故的事故原因归结于对连接处的设计不足。但同时，GCE 和连接处的供应商也兼施工方的 Havens 公司谁应该对连接处的设计变更负责,以及谁应该对整体设计负责,也是本次事故责任判定的关键点。

在调查和法庭听证过程中,有两个重要的事实被公之于众:

(1) 吊杆的制造商和施工方 Havens 在法庭证词中声称,他们就吊杆的设计变更问题与设计公司 GCE 通过电话并得到了 GCE 的同意。而在长达 26 周的庭审中,GCE 始终否认他们收到过类似电话,尽管改动的图纸上被印上了 GCE 的印章。

(2) 1979 年 10 月 14 日,在坍塌事故发生一年之前,仍在建设中的中庭屋顶发生过一次坍塌,坍塌面积超过 250 平方米。GCE 表示,他们曾经多达三次要求在现场安排设计代表,但由于经费限制都没有得到业主(皇冠中心重建公司)的同意。

最终,法庭判定 GCE 公司的人行天桥的设计不能满足建筑规范要求。这个结论是基于已经暴露的天桥部分的设计及图纸中的错误、误差和疏忽。而 GCE 公司宣称,是由于对设计含义的沟通失误造成变更设计的通过乃至最终被建造。法庭认为,GCE 公司既没有在设计阶段承担应尽的责任,也没有在中庭屋顶发生坍塌事故时进行应有的、详尽的调查。尽管供应商兼施工方 Havens 公司没有尽责审核和检验施工图纸,也没有向 GCE 公司特别标注出吊杆和箱梁的连接处的设计变化,但 GCE 公司的工程师们应该对图纸进行最终检查。GCE 公司并没有发现连接处的变更,也没有详尽调查中庭屋顶坍塌的原因,显然,他们错误地对 Havens 公司信任过度。最终,GCE 公司被判定对吊杆的设计变更负责。

1984 年 11 月 1 日,GCE 公司被判决为严重疏忽罪(gross negligence,比一般疏忽罪更严重一级,被定义为对应有责任全部疏忽),并被剥夺了设计资质。美国土木工程师协会(ASCE)也更改了相关规定:结构工程师对设计项目负全部责任。

4 基于道德和伦理的讨论

从工程伦理的角度来看,凯悦饭店事故的核心问题是:

谁应该对连接处的设计负责,为什么变更是 Havens 公司提出并最终建造的而 GCE 公司却要承担全部责任?谁被法律约束去履行这个责任?又如何履行他的责任?与这些问题相关的,是有关职业道德的要素:

(1) 隐含的或者说默认的工程师与社会之间的契约。

(2) 公共危机和基于公共危机的公众的知情同意权。

(3) 过失的定义和职业道德的标准。

(4) 另一个被公众所关注的道德问题在于救援过程中,Waeckerle 教授所代表的搜救团队对伤者做出的选择和牺牲。

Unit 7
Hotel Collapse Singapore

Teaching Guidance for Watching, Listening & Reading

Watch videos, pay attention to the *Words* and *Expressions* and related *sentences* and *paragraphs*.

On March 15, 1986, Singapore's New World Hotel *collapsed* in less than 60 *seconds*, 50 people were buried under the *rubble*, and only 17 survived. The accident was Singapore's worst disaster since World War II, like a major earthquake that *shook* the whole of Singapore. After thorough investigation, investigators found serious mistakes in the initial design of the New World hotel. The survey also changed a series of laws and regulations.

7.1 Staff in Hotel New World

Singapore's fame throughout Asia is for its order and calm efficiency. Then suddenly one Saturday morning, disaster stopped the entire country in its tracks. A six - *storey* hotel *crushed* to the ground in under 60 seconds, leading 33 people died and more than 17 buried alive beneath thousands tons of *steel*, glass and *concrete*. It has been the worst collapse in the country's history and it left the investigators to battle. Now using the advanced *computer simulations*, we revealed exactly what caused the *tragedy*. Disasters don't just happen. They are a *chain* of critical events, on arrival the *clues* and count down those final seconds from disaster.

Singapore in Asia, filled with over 4 million people, is a country with the size of New York City. It's a dynamic vibrant place that attracts people from all over the region and beyond striving to make a better life for themselves in this English - speaking island of opportunity. *Downtown* top blocks *soar* into the sky as architecture struggling to maximize the limited space. Many land *reclaimed* from *swamp* and sea. Even below ground, the city is teeming with activity. Singapore expands in the breath taking pace. Giant *tunnels* are board by team engineers from all over the world, working on the new subway system that all connects the high rise area with old and less prosperous districts, like this one——little India. Six kilometers from downtown, this area rises home to the *manual labors* who support the booming economy.

Here on Owen Road stands the Hotel New World built in 1971. The building is unremarkable, 36 *reinforced concrete columns* supporting 6 *concrete floors*. There's a car park in the basement. The *local branch* of a bank takes up *the ground floor*, above that is a popular nightclub. And a *budget hotel* occupies the rest of the building. As it's usual in this hot climate, *air - conditioning units* and *water tank* are installed on the roof. With over 24 meters high, it is small by Singapore standards, but it is one of the tallest buildings in little India. The building has led an untroubled life apart from one serious incident. In 1975, a *gas leak* poisoned 35 people. Everybody made a full recovery, and the building was given a clean bill of health.

On March 14th in 1986, 11 years later, it was a Friday night, and as usual, madam Lily Teo, the nightclub hostess, set up the first floor bar, in preparation for a busy evening, expecting the usual businessmen and couples who visited her club. She startled to see that a column on the dance floor has cracked. It was **7 pm**, she decided to inform the building's owner. Shortly afterwards, with night club open, a couple of workmen started repairing the damage. No one paid much attention to them. By **9:15 pm,** the night club was in full swing. In a staff dressing room, one of the girls checked her makeup. (Suddenly, the mirror in front of her cracked). In Singapore like most countries around the world, a broken mirror is a symbol of bad luck. But it was the end of the working week, and everyone in the night club was too busy having a good time to worry about silly *superstitions*. The next morning, On Saturday March 15th, at **8:00 am**, Cheong Cheng Guan, the assistant manager of the ground floor bank, walked through the park with his childhood sweetheart. They were married 3 months ago. But he has been working so hard, they haven't have time for their honeymoon yet.

Cheong Cheng Guan: We were happily married and we actually planed a trip to Australia and New Zealand.

As Cheong left for work, he said goodbye to his wife Leong Seey Seey, knowing it was a big day for her. She was taking an important exam.

Leong Seey Seey: I was having my final exam for diploma, Cheong and I had a plan to meet in the afternoon.

As it was Saturday, the bank was closed at lunch time. The couple were planning to spend the rest of the day walking hand in hand through Singapore's busiest streets.

Cheong Cheng Guan: I was looking forward to Saturday afternoon, after work we usually would go out.

17 people worked in the bank, it was a close friendly team. As it was common in Singapore, they took great pride in providing their customers with efficient service. The 21 year – old Christian Phua, has been working at bank for just 3 months.

Christian Phua: This was my first job in the bank, and I enjoyed it very much.

Christian has never been up to the night club on the floor above. But she was sitting almost directly below the column on the dance floor that cracked the night before.

7.2　There were Problems in the Basement Car Park

At **10:10 am**, a customer came into the bank, she was out of breath, saying there was a problem in the basement car park. *Debris* was falling down.

Christian Phua: The customer told us the bank was going to collapse.

Cheong stayed in the bank, while Christian and a colleague Sean rushed down to the basement to investigate. Although none of the cars were damaged, as they were leaving, they saw two men in overalls working on a cracked column. There was debris on the floor, but the workmen said everything was fine. Since they were in sure there was nothing serious, they went back to work in the ground floor bank. Two floors above in the hotel reception, staff saw cracks and *fissures* spreading across the wall. But down below, Cheong and Christian weren't aware of any further problems. Throughout the building, vital supporting walls and *pillars* were starting to *fracture*. The Hotel New World was on the brink of collapse.

Something extraordinary was happening within an unremarkable building in Singapore little India. Cracks were spreading throughout the walls and supporting Columns. But in the bank on the ground floor, the assistant manager Cheong is oblivious, he was wondering how his wife Seey Seey has done in her exam.

Cheong Cheng Guan: I wondered to see how she fared during exam.

7.3 An Unstoppable Collapse was Settled

At 11:22 am, further back of Cheong, Christian was finishing off the week's work when she startled by a sound, it was falling debris.

Christian Phua: I actually did not know what's happening.

At that time, she started to feel *vibrations* all around her.

Christian Phua: We could feel the floor was shaking.

At 11:26 am, the whole building was starting to shake uncontrollably. Pillars were cracking, walls were *giving way*, an unstoppable catastrophic collapse was set in motion, as shown in **Fig. 7 – 1**.

(a) (b)

Fig. 7 – 1 Unstoppable Catastrophic Collapse

Cheong Cheng Guan: It was very sudden. One instant you were working, the next instant, it was all pitch dark.

Thousand of tons of reinforced concrete rained down on the people inside. In one terrifying minute, the Hotel New World suffered total structural failure, trapping *dozens of* people. What was once a local *landmark* at the place of work was now just a huge amount of rubble, surrounded by an enormous cloud of dust visible for kilometers around. All six storeys of the building have been destroyed, transformed into a concrete tomb. Christian, Cheong and other bank staff were buried at the very bottom of the building rubble. In the pitch – black basement car park.

Cheong Cheng Guan: We couldn't see anything, we couldn't hear anything. We couldn't even see our hands.

Cheong couldn't see his colleague Christian even though she was just a meter or two away pinned down by massive concrete slabs.

Christian Phua: I was just like lying in the coffin, I couldn't move it off.

Back on the surface, it was a scene of total devastation. News crew filmed the local people *frantically* clambering onto the *wreckage*. They *furiously* manbandled rubble, desperately searching for any survivors.

7.4 Emergency Workers Started Arriving

At 11:35 am, within 8 minutes, *emergency workers* started arriving. Singapore's top *military medic* Dr. Lim Meng Kin rushed to help. He was *stunned* by the scale of this unprecedent catastrophe.

Dr. Lim Meng Kin: It has been the biggest disaster since the World War II.

The terrible news was spreading to *grief – stricken* relatives who *mingled* among the crowds of stunned *onlookers*. It was a scene of total bewilderment and rising dread.

Around noon, over half an hour into the disaster, Cheong's wife Seey Seey arrived and realized that her husband was buried somewhere inside the mountain of the rubble. She had no way of knowing whether he was dead or alive. She found her uncle who happened to be a *structure* engineer, but he could offer her no real comfort.

Leong Seey Seey: I remembered asking him what were the chances of Cheong surviving in such a collapse. He told me we should just hope for the best.

About 10 meters from Seey Seey below the rubble, her husband Cheong was fighting off the panic. But all around other survivors like his colleague Christian were aware of shouting *hysteria*.

Christian Phua: I could hear a lot of people crying. It was asking for help. Some of them were screaming and some of them were crying in fear.

Some of the survivors were badly injured. But the assistant manager Cheong was determined to keep everyone calm.

Cheong Cheng Guan: I thought it was a matter of time that people would come to rescue us.

7.5 About an Hour after the Collapse

At 12:30 pm, about an hour after the collapse, back on the surface, there was good news. The rescue team found a survivor, a tourist who was staying on the top floor of the hotel. Shortly afterwards, they found a second survivor, a local shopkeeper from hotel debris. The rescuers hastily put together a list of all those known to have been in the building when it collapsed. They could account for everyone known who has been in the hotel on the top four floors. 17 are dead and 11 have been rescued alive. The nightclub on the floor below was empty. But the ground floor bank was a different matter. Their list revealed that about up to 20 people including Cheong and Christian were working there when the building collapsed, and would be buried at the very bottom of the rubble.

7.6 12 hours after the Disaster

At 11:30 pm, 12 hours after the disaster, the rescue teams were working through the night trying to reach the bank. *Heavy machinery* held huge slabs of debris away from the top. But the rubble was very unstable and shifted dangerously. Some 10 meters below, the glare of the rescuers' art lights with vibrations shaking the wreckage increased the horror. The survivors were trapped in the dark tomb, under six thousand tons of concrete and steel. Assistant manager Cheong was just within reach one of his bank colleagues who was lying injured nearby, crushed by a *beam*. He tried desperately to push the beam away. But it was too heavy.

Cheong Cheng Guan: I felt very helpless, your colleagues were here dying and then you couldn't do anything else. It was…I felt bad.

Cheong realized his efforts were in vain.

Cheong Cheng Guan: I could smell the stench of death. It was very sad . Your collegue who has been working for so many years with you, in all of sudden, was gone.

The crushed *masonry* was preventing the circulation of air, oxygen was beginning to run out. Just a meter or two from Cheong, Christian was becoming desperate.

Christian Phua: I was thinking I was still very young, I still got a lot of things to do. And I was not married yet, I couldn't just die like that.

Back on the surface, the rescue still had 10 meters of rubble to break through. At this rate, it would take days to reach Cheong and Christian. With the oxygen getting thinner, they couldn't wait so long.

At **midnight** , half a day has passed since the collapse of Singapore's Hotel New World building. Only 11 people have been pulled alive from the rubble. Christian and Cheong were trapped deeply in the basement car park. A team of Irish engineers working on the nearby subway construction was watching the rescue work. Subway tunneler Tommy Gallagher was especially concerned that the use of heavy machinery might cause the rubble to shift and fall, potentially endangering the lives of any survivors buried at the very bottom of the mount.

Tommy Gallagher: We kept telling senior people in charge, we thought there were people alive. What you are doing here, you're gonna kill them.

They proposed a *radical* alternative, abandoning the use of *cranes* to clear the rubble from the top and instead, tunneling deeply into the foundations of the building to get the survivors out. It was a very difficult decision, But around dawn the authorities gave the green light to Tommy and his colleagues. The Irish engineer had no idea exactly where in the basement any survivors may be. After close examination of the building plans, they decided the quickest way to get into the car park was to break through *ventilating shafts* , accessible from street level. A rescue camera revealed a scene of devastation. The tunnelers had to wade through knee – deep water, and pushed fast crushed cars. To their horror, they could smell petrol. They knew a single spark could set up a *devastating* explosion. The car park was a deathtrap.

Tommy Gallagher: That's what we were worried about, any spark of machinery.

The subway engineer had an agonizing wait whilst the water and petrol were pumped out. Then they started hacking a makeshift tunnel through the debris, using small *hydraulic jack* to support the thousand tons rubble above.

At **7: 00 pm** on **Sunday** , it took 7 hours to tunnel 9 meters.

Tommy Gallagher: The ground above was very unstable, we were just taking a chance.

Then their worst fears were realized.

Tommy Gallagher: It started there – close enough.

The tunnel caved in. Tommy and his mates *scrambled out* moments before the roof collapsed. Their path was totally blocked.

At **7:30 pm** on **Sunday**, Cheong and Christian have been buried in the *stifling* claustrophobic darkness for 32 hours. They had no idea whether rescuers were trying to reach them. Undeterred by the cave – in, the Irish engineers started digging more tunnels in the dangerous building rubble, willingly risking their lives.

Dr. Lim Meng Kin: The panelists were the bravest people I've ever met in my life. And the way they went and the spirits were very rare, really.

The tunnel two and three were built from the southeast, but the subway engineers soon found their path blocked by huge slabs of debris. With tunnel one still too dangerous after the cave – in. a forth tunnel was dug northwards. It took all night. But Tommy and his mates crawled under crashed cars and dug all the way around the *exterior* basement wall.

Tommy Gallagher: We worked round the clock, we never went to bed.

7.7 Almost 2 Days Have They Been Trapped in the Rubble

At **7:00 am** on **Monday**, Cheong and Christian have been trapped in the rubble for almost 2 days. They were losing track of time.

Christian Phua: We didn't even know how many days we were trapped inside.

Cheong Cheng Guan: I was just trying to keep alive. Because I've a lot of things to live for.

Then Cheong heard a noise.

Cheong Cheng Guan: We could hear rumbling sounds from the distance.

Hoping that it was the rescue team, the bankers started tapping *frantically* on the beams that were trapping them.

Christian Phua: We were very happy, they knew where we were.

The tunnelers crawled ever close to Cheong and Christian. They felt them on the verge of reaching them. But they faced one last *obstacle*.

Tommy Gallagher: Another sudden was a meter square concrete beam that was right in front us. Our heads struck on, so they were on the other side of that.

They had no idea where in the building the beam came from. But they were desperately concerned that it helped to support the weight of the collapse building above them. Six thousand tons of rubble were balanced delicately above the heads of Christian, Cheong and now the tunnels.

It was **noon on Monday**, two days into the disaster, the rescue workers decided they had no choice except to break through the huge concrete beam. Dr. Lim joined them, ready for any emergency.

Dr. Lim Meng Kin: you didn't think of danger, there was somebody at the end of tunnel and you wanted to reach the person.

The *tunneler drills* cut through the concrete surprisingly fast. A few meters away in the darkness, Cheong heard the drilling getting louder.

Cheong Cheng Guan: They kept asking me, "Can you see a light, can you see a light?" Finally I could see a chink of light from somewhere.

Soon Cheong was able to see the rescuers just on the other side of the hole. The tunnelers were terrified that they might have cut through a major supporting beam, they felt the rubble could come crashing down at any moment. They told Cheong to stay where he was, while they were propping up the roof.

Cheong Cheng Guan: I said, "No, thank you, I'll come up".

Cheong could not bear to stay another moment in the darkness. He scrambled toward safety.

Tommy Gallagher: He just ran, I said "I don't blame you, either." Because he thought he would never come out of there alive.

After 2 days buried alive, Cheong was able to walk out into the daylight with just few *scratches*.

Cheong Cheng Guan: It was really a miracle.

Christian was buried deeper in the rubble, it took another 7 hours to rescue her. Dr. Lim led Christian in her way on the stretcher. Later, examinations revealed her only injury was a black eye.

Christian Phua: I felt just like a hero, something like that.

Cheong and his wife Seey Seey finally *reunited* at the hospital.

Leong Seey seey: It was such a moment of joy, a such great enjoy.

Cheong Cheng Guan: You just felt the life is very fragile, anytime we can be taken away from you, you just live.

Thanks to the *pioneering* use of tunnels to dig into the very deepest parts of the rubble, six more survivors were ultimately pulled alive from the wreck of Hotel New World. In all, 17 people have been saved, but 33 have lost their lives.

7.8 The Cause of the Collapse May be an Explosion

By rewinding the events of that fateful day and by going deep investigation, we can reveal what really happened to the Hotel New World. Advanced computer simulation will take us where no camera can go into the heart of the disaster zone.

News Reporter: A commission of inquiry will be appointed to seek answers to the many questions raised in the worst building disaster in Singapore.

The investigation team included Terry Hulme, a structural engineer based in Singapore. He has decades of experience in major building construction. Hulme visited the site within hours of the collapse, and was shocked by what he saw.

Terry Hulme: The first impression was that the whole collapse must have been very rapid, and the whole building was to collapse at the same time. It was clearly something unusual.

In a city like Singapore, it was everyone's worst *nightmare*. Similar buildings are everywhere and much taller structures saw skywards all across the island. The investigators needed urgently to find out why the Hotel New World collapsed. How did a six – storey reinforced concrete building collapse in fewer than 60 seconds. Until the investigators found out, the lives of millions who live and work in the similar structures throughout Singapore could be at risk. The team interviewed locals to remember that 10 years earlier a gas leak knocked out 35 people of the hotel site. Could gas have escaped again causing an explosion or worse could there have been a bomb? It was investigator Terry Hulme's first hunch. He knew that explosions left telltale signs what the experts call a distinctive signature.

Terry Hulme: The immediate effect of exposion is normally to blow out the windows. Possibly blow out the sidewall. You would see some sort of evidence of this explosion blowing the debris out or window shattered.

But the unique marks of explosion——shattered glass and crushed rubble flown hundreds of meters from the disaster site simply did not exist.

Terry Hulme: There's nothing like that.

So, if it wasn't an explosion, what could have caused the collapse?

Terry Hulme: It is really quite difficult to demolish a complete building very neatly.

7.9 The Cause of the Collapse May be the Instability of the Foundation

The team needed to investigate the disturbing possibility that the building materials were *defective*. They interviewed the Irish tunnelers and learned that some of the concrete was so soft that drills went through it like butter. Could it be that the concrete was badly mixed in the first place, *undermining* the strength of the building? With millions of lives depending on similar materials throughout the city, it was a horrific force. The team took 240 concrete core samples to one of Singapore's top *laboratories*. Of these, 80 were quickly dismissed as being unusable and unsuitable for testing. The scientists examined the remaining 160 cores to see whether the correct ratio of sand, stones and *cement* was used. Then they performed concrete strength tests on them. The cores did fracture, but only under very heavy loads. The materials they tested met internationally accepted safety standards.

Terry Hulme: It was a disappointment, because that was the easy solution, if the concrete being poor quality, you immediately find to an answer to the reason of collapse.

Hulme concluded the reason that the concrete seemed soft was because it was fractured during the violence of the collapse itself. So if it wasn't defective building materials, what else could have caused the disaster? The investigators turned to another terrifying *scenario*, one like the concrete that could have implications for the entire city. They looked at the land on which the hotel was built and discovered like much in Singapore, it used to be a swampy *flood plain*, drained from the 19th Century onwards. This was fine for the construction of small buildings, but like much of the booming city, the Hotel New World was significantly taller, bigger than anything built 100 years ago. If the reclaimed ground was unstable, could the collapse be just the beginning of an island – wide disaster? The first step for the investigators was to examine the surviving basement walls. If the foundations have moved in the unstable ground, these should show evidence of *catastrophic* cracks. Intriguingly, they found nothing. To make sure, they analyzed the composition of the soil. The team drilled deep into the ground to take samples. Leaving nothing to chance, they also tested the strength of part of the foundations that survived. Finally, after weeks of work, the results came through. Although there was evidence of some small ground movements. The team was forced to conclude that it was not enough to have brought down the building.

Terry Hulme: It became clear that they hadn't been a failure in the foundations. They were quite reasonably well constructed.

Yet another avenue closed, the team turned their gaze even further underground. Could the construction of Singapore's new subway it who weakened the building's foundations and caused it to collapse?

7.10 The Cause of the Collapse May be the Impact of the Construction of the Subway

Terry Hulme: When you build an underground railway in any town, anything that goes wrong is immediately laid to the blame for the underground railway.

Could the subway tunnelers be responsible for the disaster itself?

The investigators quickly established that the distance of two nearby subway tunnels was less than a kilometer away. What effect could this have had on the building? Subway tunnels have collapsed before. Just a year earlier in South Korea, a building fell into one. It only happened because the building was almost directly above the tunnel. But could there be an unknown mechanism by which the subway was responsible? The Singapore investigators measured the diameter of the tunnel, and calculated the ground movements that vibrations would have caused. Even taking this into account, it seemed an unlikely solution to collapse.

Professor Jonathan Wood is the world expert who studied the dynamic of building collapses for over 20 years.

Dr. Jonathan Wood: If you're within 2 diameters of the tunnel, you can expect quite an effect. Here, they were hundreds of yards away.

The team concluded that the subway tunnels were simply not close enough to be responsible for the collapse. Tommy and the tunnelers were in the clear.

Tommy Gallagher: We knew it was something that happened. A freak accident was exactly what it was.

It was now clear that collapse of the Hotel New World was not being caused by an explosion, by shoddy construction, by swampland or by subway tunneling. But buildings didn't just collapse.

Dr. Jonathan Wood: It's like a murder investigation, you must find out who is responsible.

7.11 The Cause of the Collapse May be the Micro Crack of the Pillar

The investigators knew they must be missing a vital piece of evidence. It soon arrived from a very unlikely source——The nightclub. The police interviewed the women who were working in the first floor club the night before the collapse. The nightclub hostess told them, as she was setting up the bar, a pillar on the dance floor cracked. The investigators consulted building plans, which showed this is column 26 out of a total of 36. They knew that a failure of one supporting pillar should not have brought the entire building crashing down. Then they interviewed one of the girls who was working in the night club. She told them she was checking her makeup in the dressing room, when the mirror shattered. The mirror was attached to another support pillar, column 32. The investigators now knew that a second pillar was close to failure. Next the team interviewed survivor Christian Phua. She told them on the morning of the disaster, she went down to the basement car park, and saw workmen propping wood against the column. The pillar had cracked and there was plaster on the floor. They identified this as column 30, and it was also failed. It was a crucial breakthrough. There were now eye witness' accounts that 3 columns were failing in the hours leading up to the disaster. For the first time the investigators knew that 3 vital supporting columns were stressed right to the limit. There had to be something very badly wrong with the building, but what could it be? As a last resort, the team started looking at cutting - edge laboratory research into a little - known phenomenon that could provide the answer. When concrete is stressed to be near breaking point, tiny cracks can start spreading deep within the heart of the material. Not only are these microcracks as they're called potentially deadly, but frighteningly they're invisible to the naked eye.

Dr. Jonathan Wood: You can only see them under a microscope.

7.12 Analysis of the Causes of Micro Cracks

It's impossible to know for sure whether the crucial supporting columns of the Hotel New World were suffering from this dangerous disease. But if *micro cracks* are ignored, they eventually cause the surface of pillars to fracture, greatly reducing the amount of weight they can support. At this point, the building would be on the verge of collapse. If the team was correct, like woodworm eating away *timber* from within, micro cracking would have weakened the concrete columns in the building until they were effectively rotten. But where could the extra weight that caused microcracking have come from? As the site was being cleared, the investigators saw massively heavy objects being pulled from the rubble. They examined the blueprints, to see whether these were allowed for in the original plans, and made a startling discovery. During the 15 years of its lifespan, the building owner added extra loads to the Hotel New World, which were not part of the original design. The team learned that in 1975, the bank built a steel reinforced *strongroom* weighing 22 tons on the ground floor. Then in 1978, the building owner installed two air conditioning towers, adding extra weight. In 1982, to improve the building's lackluster appearance, workmen fixed heavy-duty ceramic *glazed tiles* to the exterior walls, the weight were over 50 tons. Finally, in 1986, before the collapse, the building owner installed yet another air-conditioning tower on the roof to make conditions in the sweltering Singapore's heat more comfortable. The investigators believed all this extra weight must be too much to bear. All modern buildings are designed to support what the experts call the *dead load* and the *live load*. The dead load is the weight of the building itself. But for the investigators, the crucial calculation is the live load. The extra weight consisted of people and objects, ranging from air-conditioning towers to furniture which our building had to support. Could the building support over 100 tons of extra live load placed upon it? The answer was a big surprise.

Terry Hulme: The calculations showed that the live load was adequately supported.

Compared with the 6000 ton weight of the building itself, the live load including 100 tons of strongroom, tiles and air-conditioning units was insignificant. It should not have caused the building's 36 pillars to give way. The team was back at square one. After months of *painstaking* work, they still didn't know what happened. The team was certain they must have missed some vital clue that would enable them to understand why Singapore's Hotel New World collapsed. They returned to the blueprints, and poured over photos of damage columns again and again. Clutching at straws, they double-checkd the calculations made by the draftsman for the weight of the building itself, the dead load. What they found was a revelation, a fatal mistake that meant for the entire 15 years after its construction the building was on the verge of catastrophe.

Terry Hulme: The amazing thing was basically, that this draftsman allowed for the live load and even forgot the dead load.

It was an astonishing discovery, a schoolboy mistake. The calculations made by the draftsman were so badly wrong that key columns, such like 26 and 32, could not even support the weight of the building itself. Even without all the extra live load, including the air conditioning units, heavy-duty glazed tiles and bank strongroom, for 15 years, many of the pillars that were supporting the building had been right at the limit of their strength. The collapse was just a matter of time.

Dr. Jonathan Wood: All the columns were very near failure, being sitting close to the edge of collapse for 15 years.

The investigators finally understood why the disaster happened.

Now by rewinding the events leading up to that fateful day, and by following the process of the extensive investigation, we can finally reveal what really happened to the Hotel New World.

7.13　Conclusions of Investigators

On Friday of March 14th, 1986, for 15 years hidden from human eyes, deadly micro cracks were spreading deep within the concrete columns of the Hotel New World. At 7:00 pm, 16 hours before the collapse, column 26 cracked in the nightclub. It was weakening and started to pass its load to surrounding pillars. About 13 hours before the disaster, column 32 couldn't support the extra loads and started to fracture. This caused the mirror attached to it to smash. Just over one hour from the disaster, load from two nightclub columns was being transformed to column 30. It started cracking, this time was in the basement car park. 4 minutes before disaster, above Christian, column 26 started to collapse, causing debris to pull down, and vibrations to spread through out the bank. With 3 columns' collapse, the building reached the point of no return, an unstoppable chain reaction was set in motion.

Dr. Jonathan Wood: If you call it a pack of cards, you've got to push down one card and all the other cards follow soon.

The entire building crashed to the ground in under 60 seconds, killing 33 people. Tragically, the investigators now knew there were plenty of warnings that the hotel was about to collapse. If the building owner had told the authorities that 3 columns were cracking, instead of ordering workmen to patch them up. Disaster could easily have been averted.

Dr. Jonathan Wood: I believe any structure engineer would've immediately said," you need to evacuate the building".

33 lives could have been saved. Survivors Cheong and Christian will always remember their colleagues who died in the Hotel New World collapse.

Christian Phua: I do miss all my colleagues, they are the best ever have.

But they will also never forget the courage of the subway tunnelers.

Cheong Cheng Guan: It was volunteering, and they had dug through debris which might just fall on them at any time.

Christian Phua: Without them, I'll not be here, so I would like to thank them very much on this.

Tommy Gallagher and subway tunnelers were awarded one of Singapore's highest Peacetime honors——the *conspicuous gallantry medal*. Lessons have been learned from the tragedy. Singapore completely *overhauled* its building regulations, to ensure that all dead and live load calculations made by architects would in future be independently checked. A crucial extra safeguard has been adopted by many other countries. Like many cities, Singapore continues to build ever upwards. Mercifully, building collapses are rare events, both in Singapore and throughout the world. But when they happen, the courage and dignity of ordinary people shine through.

Words and Expressions

collapse 倒塌，瓦解
second 秒
rubble 碎石，碎砖
shake 摇晃，震惊
storey 楼层
crush 压碎，粉碎
steel 钢筋，钢结构
concrete 混凝土
tragedy 悲剧，灾难
chain 链
clue 线索，迹象
downtown 市中心的
soar 高耸
reclaim 回收再利用
swamp 沼泽
tunnel 隧道，地下通道
column 柱
superstition 迷信，迷信观点
debris 碎片，残骸
fissure 裂缝，裂痕
pillar 柱子，栋梁
fracture 破裂，断裂
vibration 振动，颤动
landmark 地标
structure 结构，建筑物
frantical 疯狂的，狂暴的
wreckage 残骸，受灾地点
furiously 猛烈地，狂暴地
stunned 受惊的，震惊的
grief-stricken 极度悲伤的
mingle 混合，使混合
onlooker 观众
hysteria 歇斯底里
beam 梁
masonry 砌体，石造建筑
radical 激进的，极端的

crane 吊车，起重机
devastating 毁灭性的
stifling 令人窒息的
exterior 外部的，表面的
obstacle 障碍，干扰
scratch 划痕
reunit 重聚，使再结合
pioneering 首创的，先驱的
nightmare
defective 有缺陷的，次品
undermine 暗中破坏，逐渐损坏
laboratory 实验室
cement 水泥
scenario 方案，情节
catastrophic 灾难的，悲惨的
timber 木材，木料
strongroom 保险柜
painstaking 刻苦的，下苦功的
overhaul 彻底检查，大修
building engineer 建筑工程师
computer simulation 计算机模拟
manual labor 体力劳动，手工
reinforced concrete 钢筋混凝土
concrete floor 混凝土楼层
local branch 本地分行
the ground floor 一楼
budget hotel 经济型酒店
air-conditioning unit 空调设备
water tank 水箱
gas leak 漏气，气体泄露
give way 倒塌
dozens of 许多，几十
emergency worker 救援人员，急救人员
military medic 军事医生
heavy machinery 重型机械

ventilating shaft 通风井	micro crack 微裂缝
hydraulic jack	glazed tile 琉璃瓦
液压千斤顶，液压起重器	dead load 恒载，静负荷
scramble out 争先恐后出来	live load 活载，动荷载
tunneler drill 隧道掘进机钻头	conspicuous gallantry medal
flood plain 满滩，泛滥平原	杰出英勇勋章

Translation Examples

［1］ Throughout the building, vital supporting walls and pillars were starting to fracture. The Hotel New World was on the brink of collapse.

整栋大楼最重要的支撑墙面和柱子开始破裂，新世界酒店随时可能倒塌。

［2］ Thousand of tons of reinforced concrete rained down on the people inside. In one terrifying minute, the Hotel New World suffered total structural failure, trapping dozens of people.

大楼几千吨的钢筋混凝土，大批掉落在人们的身上。在这恐怖的1分钟里，新世界酒店的结构完全破坏，有数十人困在里面。

［3］ They proposed a radical alternative, abandoning the use of the cranes to clear the rubble from the top and instead, tunneling deeply into the foundations of the building to get the survivors out. It's a very difficult decision.

他们提出一个极端的做法，不再使用起重机从顶端清除瓦砾，反而挖地道通往大楼的地基，把生还者救出来，这是一个很困难的决定。

［4］ The team needed to investigate the disturbing possibility that the building materials were defective. They interviewed the Irish tunnelers and learned that some of the concrete was so soft that drills went through it like butter. Could it be that the concrete was badly mixed in the first place, undermining the strength of the building?

调查小组必须研究一个令人不安的可能，即建材可能有缺陷。他们咨询了爱尔兰的隧道工程人员，发现某些混凝土非常的松软，钻孔机一钻就破。会不会是混凝土一开始就没有浇筑好，逐渐损害、破坏了大楼的强度？

［5］ But if micro cracks are ignored, they eventually cause the surface of pillars to fracture, greatly reducing the amount weight they can support.

但如果忽视了微裂缝，最后可能使柱子表面破裂，大幅度降低柱子能够承载的重量。

［6］ On Friday of March 14th, 1986, for 15 years hidden from human eyes, deadly micro cracks were spreading deep within the concrete columns of the Hotel New World.

1986年3月14日星期五，15年来在肉眼看不见的地方，致命的微裂缝在新世界酒店混凝土支柱内部的深处不断蔓延。

［7］ Lessons have been learned from the tragedy. Singapore completely overhauled its building regulations, to ensure the all dead and live load calculations made by architects would in future be independently checked.

我们必须从悲剧中吸取教训。新加坡全面检查建筑规章，确保建筑师所做的所有静荷载和动荷载的计算，将来都要经过独立的审核。

Activities—Discussion, Speaking & Writing

Presentation

Group: 5 to 7 members

10 minutes per group (Each member should cover your part at least one or two minutes).

Clearly deliver your points of the following questions to audiences.

NEED practice (individually and together)!!

Gesture and eye contact.

Smile is always KEY!! Cover your nervousness!!

Questions for Discussion and presentation

A disaster on this scale raises several crucial questions:

1. Could it be that the concrete was badly mixed in the first place, undermining the strength of the building?

2. Could the construction of Singapore's new subway weaken the building's foundations and cause it collapse?

3. Could subway tunnels be responsible for the disaster itself? What effect could it have on this building?

4. Where could the extra weight that causes the micro cracking have come from?

5. Could the building support 100 tons of extra live loads?

Writing

Read **Hotel Collapse Singapore**, write a short essay in English in groups of six members which contains the following information.

新世界饭店倒塌事故

1986年3月15日,新加坡的6层新世界酒店在不到60秒时间内轰然倒塌,50人被埋在碎石下,最后只有17人生还。这起事故是新加坡在二战后发生的最严重灾难,像一场大地震,震动了整个新加坡,灾难现场如图1所示。调查小组经过几个月辛苦的调查,确定不是因为瓦斯爆炸、施工质量不佳、建筑材料质量不合格、沼泽地填筑地基不稳定,也不是因为地铁隧道影响。调查小组确信他们错失了某个重要线索,才无法解开新加坡的新世界酒店倒塌的原因。在令人恐怖的1分钟内,整个新世界宾馆大楼崩塌了,钢筋混凝土废渣雨点般地向人们砸下来。这场灾难造成了33人死亡、17人被活活困在数千吨重的钢筋、混凝土和玻璃渣下面。这是新加坡历史上最严重、也最令调查人员困惑不解的倒塌事件。

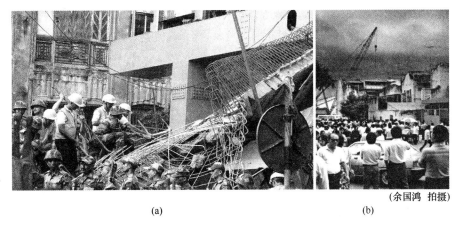

(余国鸿 拍摄)

图 1　新世界酒店倒塌现场

一、极端救援方法：开挖隧道直达倒塌大楼底部

1986 年 3 月 16 日，星期日，午夜 12 时。大约半天之前，新世界宾馆大楼以惊人的速度发生了整体倒塌，数十人被埋在废墟中生死未卜。经过 12 个多小时的奋力救援，仅有 11 人被活着救出废墟，救援工作进展缓慢，尚有幸存者被埋在漆黑的地下停车场里，按照正常速度，救援人员要在几天后才能到达他们的深度。因为塌下来的石块阻止了空气的流通，周围的氧气越来越稀薄，他们随时可能丧命。

1986 年 3 月 17 日，星期一，中午时分。隧道钻孔机以惊人的速度切割着新世界宾馆大楼的混凝土废墟。在几米开外的黑暗中，幸存者之一张清源听到了钻孔机的声音。张清源回忆说："他们（救援人员）不停问我，你看得见光吗？看得见吗？最终，我看见一束光从某处射了进来。这真是一个奇迹！"在被埋 2 天之后，带着身上的一些擦伤，他终于重见天日。另一名幸存者 Christian 被埋在更深的碎块下面，搜救队员又用了 7 个小时才将隧道挖到她身边，把她解救出来。

Christian 和张清源都是一楼的工商银行分行的员工，大楼倒塌后，他们被埋在最底下的地下停车场里动弹不得，是在附近修建地铁的爱尔兰工程师们开挖的隧道救了他们的命。这些工程师同当地民众一样关注着救援行动的进展情况，其中一名叫 Tommy Gallagher 的工程师担心大量重型机械会让坍塌的碎石再次动摇坠落，从而危及依然被埋在废墟下面的幸存者。于是他向现场负责营救的总指挥提出了一个看似极端的方法——放弃使用起重机清理倒塌物，而是开挖隧道直达大楼底部的停车场，救出幸存者。这是一个困难的决定，不过总指挥最终采纳了 Tommy 的建议。最终，这次事故共有 17 人获救，同时 33 人不幸丧生。

二、持续两天的多处开裂

现在，退回到事故发生的前一天，按照时间顺序让这场灾难情景再现。

1986 年 3 月 14 日，星期五，晚上 7 点。海神环球夜总会舞池里的一根柱子（在酒店的设计平面图上编号为 26）突然开裂，正在一边擦拭酒杯、准备接客营业的老板娘张莉莉（音译）顿时吓了一大跳，她于是马上通知了大楼的业主。几名工人很快便将裂缝修补好，夜总会照常营业，没有人留心这起意外。

1986 年 3 月 14 日，星期五，晚上 9 点 15 分。夜总会的员工更衣室里，一名女子正在对着镜子化妆，门外是喧闹的音乐声和人声，舞池里气氛达到了高潮，她面前的镜子一角毫无征兆地碎开，不过碎片没有掉落或飞溅，她也没有受伤。那块镜子紧连着另一根编号为 32 的支柱。虽然和世界上很多国家的风俗一样，在新加坡，镜子碎裂也是一种凶兆。不过，对于沉浸在周末轻松气氛中忙于享乐的人们来说，这种无聊的迷信令人无暇顾及。这名女子很快整理好妆容，如往常一样进入舞池，在客人之间周旋服务。这一天有惊无险地度过，没有人在意连续发生的两起小小意外。

1986 年 3 月 15 日，星期六，早晨 8 点。位于一楼的工商银行分行的副经理张清源正和新婚妻子道别，准备上班。因为那天是星期六，银行中午就打烊，两人相约下班后去闹市区逛逛。张清源上班的分行共有 17 名员工，相互之间的感情都很好，平日里为顾客提供新加坡引以为傲的高效服务。员工之一 Christian Phua 虽然来分行才 3 个月，却已经十分喜爱这份工作。Christian 从没去过楼上的夜总会，但她工作的位置就在舞池前一晚裂开柱子的正下方。

1986 年 3 月 15 日，星期六，上午 10 点 10 分。一名女顾客突然气喘吁吁地跑到银行，她说，地下停车场突发状况了：有碎石头往下掉。她告诉 Christian：“大楼要倒塌了！”张清源留守分行，同时 Christian 和一名同事冲到地下室去查探情况。虽然他们当时没有发现任何车辆受损，但看到两名工人正在修补一根裂开的柱子（编号为 30），地上落了一堆碎渣。但工人们告诉他们，一切正常，没什么好担心的。得到确认的两人放心地回到一楼继续工作。就在此时，三楼宾馆接待处的员工们发现，墙上出现了裂缝和裂纹，并迅速蔓延到整面墙壁。但没有人知道更严重的问题正在威胁楼里人的生命，大楼最重要的支承墙面和柱子开始皲裂，整栋大楼已经摇摇欲坠，随时可能倒塌。

三、宾馆大楼整体倒塌，60 秒内生死两重天

1986 年 3 月 15 日，星期六，上午 11 点 22 分。Christian 正坐在自己的位子上总结一周来的业务情况，却被一个声响吓了一跳。"我其实不知道到底发生了什么事。"她回忆说。那其实是碎石掉落的声音，很快，几分钟后她就感觉到周围的一切都在振动。

1986 年 3 月 15 日，星期六，上午 11 点 26 分。这时整栋大楼都在失控摇晃了。柱子断裂，墙体坍塌，在不到 1 分钟的时间内大楼倒了下来，一场不可遏制的灾难降临。"事情发生得非常突然，前一秒还在工作，下一秒就眼前一片漆黑。我们什么都看不见，甚至连自己的手都看不见。"和 Christian 还有其他分行员工一起被埋在最下层的张清源说。动弹不得的 Christian 说自己当时好像在棺材里一样。在这骇人的短短 1 分钟里，宾馆大楼的建筑结构完全遭到破坏，数十人被困在重达几千吨的钢筋混凝土下面。这座曾被视为当地地标的大楼，如今只剩下了大堆的瓦砾。大量弥漫的烟尘从几公里外都能看到。六层楼高的建筑物顷刻间被摧毁，变成了一座混凝土的坟墓。

1986 年 3 月 15 日，星期六，上午 11 时 35 分。灾难发生约十分钟后，急救人员陆续抵达。新加坡首席军医林明建（音译）博士也来到了现场，震惊于灾难的规模之大，他回忆道："这是二战以来新加坡最大的灾难。"一辆辆巨大的起重机从岛屿各地汇集到灾

难现场，开始清理酒店的废墟，展开搜救行动。

四、艰难的调查最终揭开谜底

事后的调查就好像在解一个难破的迷局，类似的楼房在当地随处可见，为什么唯独这幢大楼发生了瞬间坍塌？建造 15 年来从未发生过结构问题的钢筋混凝土建筑，到底是如何在 60 秒内倒塌的？谁又该为此负责？几百万民众迫切需要知道灾难为什么会发生。

新加坡土木工程师 Terry Hulme 参加了相关部门组成的调查委员会，奉命解答因这一灾难而产生的众多问题。倒塌事件发生几小时后，Terry 查看了现场。他认为倒塌是在很短的 1 分钟内发生的，整幢建筑几乎是同时倒下来的，这很不同寻常。调查组先后进行了爆炸、建筑材料质量问题、地基不稳或受损等事故原因的推测，但最终都被一一否定了。要找到真正的原因似乎有些困难。

但没有建筑物会无故倒塌。谜底就藏在大楼倒塌前发生开裂的柱子里。通过询问事故幸存者，调查组证实了共有 3 根支柱（编号分别为 26、32、30）在事故发生前的几个小时里陆续开裂，由此推测，3 根支柱已经濒临支撑极限，这说明大楼一定出现了非常严重的问题。但它的问题是什么呢？

调查组通过查询尖端实验室的研究成果，得知当混凝土受压达到极限时，肉眼无法察觉的微裂缝就会在建材的深处扩散，新世界宾馆大楼的关键支承柱有没有出现这一致命缺陷不得而知，但如果忽视了这些只有用显微镜才看得到的细微裂缝，最终的确会导致支柱表面形成破裂，从而大大减少柱子所能支撑的总重量。如同木蛀虫从里面蛀空木料一样，微裂缝会慢慢削弱混凝土柱子的承重能力，直到完全不堪一击。假使微裂缝是导致支柱集体崩溃的原因，那么造成微裂缝的附加重量又是哪来的呢？这是调查组接下来要解开的另一个谜题。

五、15 年的隐患最终酿成悲剧的 60 秒

按照设计，所有的现代建筑都必须符合"静荷载"和"动荷载"的有关标准。静荷载是指建筑本身的重量，一开始，调查人员重点要考虑的是"动荷载"，也就是额外增加的人和物的重量，其中包括空调设备，以及建筑物里必备的一些家具。新世界宾馆大楼的动荷载是否超标呢？答案是否定的。相比宾馆大楼自身约 6000 吨的"静荷载"，总重仅 100 多吨的保险库、瓷砖和空调设备之类的动荷载根本不值一提，它们不是导致大楼 36 根柱子集体崩塌的原因。

调查人员再次翻阅了平面图，并且一遍又一遍地查看受损柱子的照片，又重复检查了制图员计算的大楼自身重量，也就是"静荷载"。结果，他们有了突破性的发现。调查员 Terry 说，万万没想到，这名制图员算进了动荷载，却完全忘了"静荷载"，这是个令人震惊的发现，一个学生才会犯下这样的错误。这名制图员错得离谱，以至于像 26 号和 32 号这样的重要支柱，根本无法承受大楼自身的重量。也就是说，15 年来，不为人察觉的微裂缝一直在新世界宾馆大楼的混凝土柱子内部延伸，支撑大楼的许多柱子也早已濒临承受的极限，大楼随时都可能化为废墟。

1986 年 3 月 14 日 19 点，灾难发生前 16 小时，夜总会里的 26 号柱子开裂。它的承重力减弱，于是部分负载被转移给相邻支柱。灾难发生前 13 小时，32 号柱子由于无法支撑

附加的重量，开始断裂，这导致紧挨着它的镜子破碎。灾难发生前 1 小时，夜总会里那两根柱子的负载被转移给了 30 号柱子。于是，这根柱子出现裂缝，致使大量的碎块落在地下停车场里。灾难发生前 4 分钟，位于 Christian 头顶的 26 号柱子开始坍塌，一些碎渣掉落下来，随后，银行里的物体开始摇晃。随着 3 根柱子的崩溃，整幢建筑摇摇欲坠。很快，它就像多米诺骨牌一样轰然倒塌。新世界宾馆大楼在不足 60 秒内变为了一堆废墟，致使 33 人丧命。

六、新世界饭店悲剧给我们的启示

可悲的是调查人员现在知道，事故前有许多征兆显示大楼即将崩溃，如果业主通知当局有三根柱子正在破裂，而不是叫工人把裂痕遮住，或许可以轻易避免灾难发生。相信任何结构工程师看到后，都会表示必须马上进行疏散，避免这样白白葬送 33 条性命。

生还者张清源和 Christian，永远记得他们在新世界酒店倒塌事件死难的同事。但他们也会永远记得隧道工程人员的勇气，他们主动请缨上阵，他们要贯穿瓦砾，随时可能被瓦砾压垮。没有隧道救援人员，就没有今天的幸存者。Tommy Gallagher 和隧道工程人员，获得新加坡承平时代最高荣誉——英勇勋章。

60 秒的惨烈，15 年的隐患。我们必须从悲剧中得到教训。悲剧过后，新加坡全面检视建筑规章，全面修改了建筑工程的规章制度，确保建筑师所做的所有静荷载和动荷载的计算，将来都要经过独立的审核。全球许多其他国家已经采用了这个重要的附加安全措施，和许多城市一样新加坡的房子还是越盖越高，但幸好建筑物很少倒塌，无论是在新加坡还是世界各地。不过一旦悲剧发生，更能展现出平凡人的勇气和高贵。

Unit 8
King's Cross Fire

Teaching Guidance for Watching, Listening & Reading

Watch videos, pay attention to the *Words* and *Expressions* and related **sentences** and **paragraphs**.

In London's King's Cross station, a small blaze suddenly erupted into the roasted inferno, spitting out fire like a *flamethrower* in a minute. What originally started fire? What fueled its growth?

8.1 Rush Hour on the City Center

It was the Christmas shopping season in London and the *rush hour* is in *full swing*. Commuters and shoppers streamed for the King's Cross, the city's busiest rail interchange. Suddenly, a deadly wall of flame rolled through the packed station, which killed 31 people. The world oldest *subway system* suffered the deadly fire in its history. Now using the advanced *computer simulations*, we revealed exactly what caused the *tragedy*. Disasters don't just happen. There are a chain of critical events, on arrival the clues and count down those final seconds from disaster.

Europe, England, London

King's Cross station is Britain's busiest rail heart. It connects the ground trains from the north with the city's 418 kilometers of subway network, unofficially known to Londoners as the *tube*. Five tube lines serve eight platforms in King's Cross. The deepest is 27 meters underground. A quarter of million passengers *surge* through the station's *subterranean labyrinth passages* every day.

On **Wednesday of November 18th, 1987, at 7:18 pm,** It was still rush hour and the city center was busier than ever as the Christmas shopping season getting to gear.

For one man, the rush and the bustle of London was his new experience. 20 year-old Daemonn Brody left his home in Scotland just 5 days ago to start a new IT job and a new life in the big city. He couldn't believe his luck.

Daemonn Brody: I always want to live in London. For some of the 20 year-old, the dream comes true.

Brody was eager to *soak up* London sights. Tonight (November 18th) he was off to visit the famous Christmas light in Regent Street. The trip must be taken on the tube. But Londoners knew the Christmas in the capital wasn't all peace and good will.

Four years ago, the Harrod Department Store was suffered a bomb attack. The blast happened eight days before Christmas and killed 6 people. Now intelligence expert felt terrorists may once again target London. But nobody knew where terrorists might strike.

At **7:25 pm**, 27 meters underground of the King's Cross, passengers streamed off trains and onto escalators, taking them up to the ticket hall and the streets of London. On the Piccadilly Line escalators, several commuters spotted someone in blue boiler suit, *descending into* the *manhole* nearby. Workermen were familiar sights in wooden escalator of the aging *dilapidated tube sys* *tem*. Two minutes later, electronics engineer Philip Squire joined the up escalator. He headed up to watch the street, obeying the tube's act by standing on the right hand side.

Philip Squire: As I was standing there, I was looking around as you do. And I happened to look down.

He caught sight of something odd, a light that underneath the wooden stairs.

Philip Squire: I saw it was very bright and very intense.

When he reached the top, he sought the *underground worker*. Then they looked back to the spot, which was half way down of the escalator where Squire saw the glow. But there was no sight on it.

Philip Squire: I felt a bit silly, because you couldn't see anything. To be honest, I felt stupid.

Philip Squire went on his way, embarrassed to have caused the fuss.

8.2 A Small Flame on One of the Wooden Steps

At **7:30 pm**, a couple minutes later, a man hit the *emergency button* on the Piccadilly line escalator. A British transform policeman on routine patrol went to investigate. He saw a small flame on one of the wooden steps, as shown in Fig. 8-1. It was not a huge surprise that fire was a fact life from aging *infrastructure* of the tube. There have been over 400 in the previous 3 decades. The policeman headed back up to the ticket hall, where he could call London *fire brigade* on his radio. Down below, staff followed routine practice, taping off the Piccadilly Line escalator to divert passengers away from the tiny fire.

At **7:39 pm**, 9 minutes after the fire alert, the police decided they should evacuate station as the precaution. Three kilometers away in the city's Soho District, the firefighter of Red Watch responded to the call, and headed for the King's Cross. Soho was the busiest fire district in London, many of men have fought hundreds of blazes.

8.3 Red Watch Arrived

At **7:43 pm**, red watch arrived at the King's Cross ticket hall. The ranking officer was 45 year-old fire chief Colin Tensile. Tensile is married with 2 daughters and served 23 years in the London fire brigade. He dreamt of retiring early and buying a rural *smallholding* in France. Tensile is a dedicated and demanding leader, who commanded respect and *loyalty* of his mass. One of them is Robert Moulton, who served with the man he called "the governor" for four years. Tensile took Moulton with him to assess the state of the fire. A *blaze* was enough about the size of camp-fire, but neither of them was worried. Sixteen meters below, British transport police continued to clear hundreds of passengers from the station. One of the police officers was the 27 year-old Richard Kukielka. After 5 years of this job, *evacuation* procedure was like the second nature.

Richard Kukielka: I just managed to get the people out of the station.

Kukielka directed the passengers onto the Victoria Line escalator, which was the only one still operating. It ran parallel to the Piccadilly Line escalator and the *emergency exit* was in the same place. The main tickets hall was just below the street level. Heading up to the ticket hall with the crowd was London new comer Daemonn Brody. He was on his way to see the Christmas lights, but his trip had just been canceled.

Daemonn Brody: There was no panic, no rushing, no pushing, it was just normal.

8.4 The Flame was about a Meter and Half High

At 7:44 pm, the fire had been burning on the wooden steps for almost 15 minutes and the flame was now about a meter and half high, as shown in Fig. 8-2. The blaze was 20 meters from police officer Kukielka at the bottom of Victoria Line escalator. *Firefighter* chief Colin Tensile was in the ticket hall. Sightseer Daemonn Brody was traveling up the adjacent Victoria Line escalator. Brody reached the top of the escalator and stepped into the ticket hall. A moment later, a huge jet fire *erupted*.

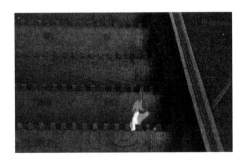

Fig. 8-1　A Small Flame on Wooden Elevator

Fig. 8-2　Flame was about 1.5m High

Daemonn Brody: It was like an explosion happening in slow motion.

Daemonn Brody took the first hit, the jet flame ate his leg and lower body.

Daemonn Brody: My life served ended at that point.

In London's King's Cross station, a small blaze suddenly erupted into the roasted inferno, spitting out fire like a flamethrower, engulfing *dozens of* commuters in a ball of fire. One of them was the 20 year-old sightseer Daemonn Brody.

Daemonn Brody: I knew that particularly the back of me was all on fire quite badly and I knew my legs were also like that.

Now big black smoke poured out the station exit into the street above. Here firefighter Robert Moulton was getting the hose ready to deal with what was just minutes before, a tiny escalator fire.

Robert Moulton: It was terrific roar and *horrible* scream sounds for a long time.

He was horrified, he knew that ticket hall was full of people moments before. Among them was his boss, the fire chief Colin Tensile. News crews, captured the scene that Moulton and his fellow firefighters grasped the *oxygen tanks*. Ignoring the danger, they descended to the burning ticket hall to search for survivors.

Robert Moulton: We only thought we just try to find a way down to the fire to drag anyone out.

Caught in the inferno, Daemonn Brody rolled over and over, tried to put the flame out. Eventually he succeeded. He was badly burnt but now faced new danger——deathly choking smoke.

Robert Moulton: There was absolutely hot and I could feel burning through throat and lungs.

Moments after the fire erupted, from the bottom of Victoria Line escalator, young police officer Richard Kukielka could see a horrifying sight.

Richard kukielka: When I looked up the escalators, what I saw was darkness.

He realized when he sent passengers up to the escalator, he unwittingly sent many of them straight into the inferno. But the blazing ticket hall was the only way out to the streets. Thinking fast, police and station staff flagged down passing trains and bundled stranded passengers out of danger. But firefighter Robert Moulton battled on through the ticket hall, searching for survivors including his boss Colin Tensile. He was astonished by the ferosity of the fire.

Robert Moulton: It was tremendously hot, seriously hot, you couldn't believe it.

Then as he shone his torch through the smoke, he saw something. It was a white *helmet*, his heart sank. He knew the white helmet was only worn by the fire chief.

Robert Moulton: I pushed my mask above, and I shone the torch in. I could see it was white, and I just knew it was Colin.

His boss wasn't moving, and even if he's alive, Moulton knew Tensile's lungs would be full of *toxic* smoke. He had to get him into the open air quickly. Deep underground, the police thought they got everyone out. But Richard Kukielka heard his colleague find a passenger slumped at the bottom of the escalators.

Richard Kukielka: Some hair was burnt and his head and his hands were badly burnt. He was in lots of pain, screaming and shouting.

The man needed urgent medical attention, but he was too badly burnt to evacuate by train. The young policeman feared they were trapped, then he remembered there was another way out up to the street. They dragged the man to the medal gate leading to the exit. But when they reached the gate, they discovered to their horror that somebody had locked them.

8.5 Open the Gate. Hello!

Richard Kukielka: The only situation commanded us was trying to kick the *iron* gate down.

He yanked frantically on the gate but it wouldn't *budge*. The young policeman couldn't radio for help as his walkie‑talkie wouldn't work underground. It was a crashing moment. And the fume from the fire ranging above, were starting to spread for all of the station.

Richard Kukielka: Our eyes were hurting and chest was hurting, and we couldn't breathe probably.

Upon the blazing ticket hall, Moulton was dragging fire chief Tensile out of the inferno. What he didn't know was that only meters away, sightseer Daemonn Brody was fighting for his life. His lungs were screaming and he desperately wanted to sleep.

Daemonn Brody: I thought now just, I just go to sleep, I hope when I wake up, everything will be over.

Dozens of people in the blazing ticket hall were choking to deadly toxic fumes. But in the streets above, news crews captured a desperate scene, firefighters were staggering out, being beaten by the *intense* heat and flames. These men knew that any hope of saving people that trapped just 5 meters below was disappearing fast.

At **7:52 pm**, seven minutes since the fire engulfed the King's Cross station, trapping commuters below ground, there was scene of complete chaos in the streets above the station. Fire fighter Robert Moulton frantically gave his boss Colin Tensile mouth to mouth to get life-saving oxygen into his lungs.

Robert Moulton: The very first time when I blew breath into, he got a cough. And I said, "it seems good", we really sent two.

And encouraged by the signs of life, Moulton redoubled his efforts. Then the paramedics rushed his boss to hospital. 16 meters underground, in one of the station labyrinth passages, 27-year-old policemen Richard Kukielka, was trying to take the badly injured passenger to safety, before the fire spread further. But he was trapped behind a locked gate and whatever he tried, the gate just wouldn't budge. Then a cleaner appeared on the other side. The cleaner had no clue that inferno was ranging above. But now he proved a life saver, he had the key to the exit. 32 minutes after the fire erupted, Kukielka rushed the wounded man out of the station to safety. Up in the blazing ticket hall, several meters from the escalator, with last breath left, sightseer Daemonn Brody was sleeping away. Then the 20 year-old was grieved by the sense of injustice.

Daemonn Brody: I only thought this wasn't right. And I wouldn't mind dying again in a week, and I wouldn't mind it happening in any other way, but I just didn't want to be here and now.

Summing up the last of his strain, he started to *crawl* blindly through the smoke, desperately searching for some way out. The ground was fiercely hot, making every meter he crawled miserable. Then through the smoke, Brody made out the stairway. He began to struggle up with inch by inch. Suddenly through the *gloom*, he saw a hand reaching toward him. A stranger bent down and started to haul Brody up the stairs out of the nightmare.

Daemonn Brody: When I got up to the top of the stairs, it was amazing.

At the King's Cross, the firefighter just controlled the blazing enough to adventure below. But in the wrecked ticket hall, all they discovered were dead bodies. Then Robert Moulton and his fellow firefighters heard the terrible news. Their boss fire chief Colin Tensile was died in the hospital of smoke *inhalation*.

Robert Moulton: We were all shocked and sorry, it was a bit *paralytic* really.

It was not until six hours later, at **1:46 the next morning** that the fire was finally contained. The inferno has killed 31 people and seriously injured another 20.

(News from One O'clock News: Good afternoon, the cause of Wednesday night fire of King's Cross underground station was still not known. The youngest *victim* is only 7 years old.)

Relatives of victims and survivors tried to come to turn with the *traumatic* experiences. Among them was 32 year-old Steve Hanson.

Steve Hanson: I have 3 children and a lovely wife. My youngest child is only six years old. That's what kept me going. I have to live for them.

The tragedy horrified the nation, uniting royalty and British public in grief. British Prime Minister Margaret Thatcher visited the site, who was visibly shocked by the devil station. But the causes of this disaster were a complete mystery. More than 2,000,000 people use London's tube system every day. If the Prime Minister shut it down, London would bring to a halt. Thatcher set up an inquiry and ordered them to find answers. What triggered the blaze, and what transformed the minor fire into a deadly jet flame in seconds?

8.6 Whether It was an Arson or a Terrorist Attack?

Now By rewinding the events back the whole day, and by going deeply into investigation, we can reveal what really happened in the King's Cross inferno. Advanced computer simulation will take us where no camera can't go, into the heart of the disaster zone.

The government appointed an investigation team of 21 experts from 5 different agencies. Engineer David Shillito was one of the leading investigators. Shillito has been an international expert in major fire disasters for 34 years. In the burnt *wreckage* in the King's Cross, shillito looked for the clues to the fire spread.

David Shillito: The burning pattern was fundamentally important, the shapes of the burning on the escalator, the way that smoke had blackened the *underground structures* showed it was a suddenly bursting form of major fire.

It was immediately clear to him that it was not ordinary fire. But he needed more evidence about how it behaved. Within hours, police officers started to *track down* anyone who might had witnessed the blaze. They interviewed hundreds of people. Several described how the jet flame shot up to the escalator and into the ticket hall, like a *blowtorch* or a flamethrower.

Steve Hanson: It was not an emerge situation but suddenly it was black inferno hell.

The sheer force of the inferno suggested a disturbing possibility——a terrorist attack. The terrorist has bombed Londoners before, is this the start of terrific new chapter in their *terror campaign*, targeting travelers on London's tube? Within hours of this disaster, police officers made a *chilling* find, three eye witnesses recalled seeing a man in a blue boiler suit minutes before the blazes. He was by the Piccadilly Line escalator where the fire started. Two of them saw him went down in the nearby manhole. Where did that lead? Two managers told police the conduit went straight to the void underneath the escalator. But none of the workers on duty that day fit the description of the man. Was it the terrorists planting an explode device under the escalator? The investigation team dropped in David Halliday, an forensics scientist who is investigating suspicious blazes. In the sealed station, technicians stripped down and examined the wreckage specialized of the burnt out escalator for clues. Half way down the escalator, they found classic pattern of scorching, that indicated the fire's origin point. It was underneath the escalator, close to the running track. But they could not find bomb creator here, or anywhere on the escalator. Nor was the metal work scattered, which was the sign of explosion.

David Halliday: All of the *steel work* and structure of the escalator was in fact generally intact. Where has distorted was clear to be distorted because of heat not the explosion.

The investigation team ruled out the bomb was the origin of the King's Cross fire. But they were still suspicious that the speed of the fire spread could be explained by another *criminal act*, arson.

Now Halliday was on the hunt for traces of *accelerant*, like petrol, the calling card of the *arsonist*. They particularly examined the area around the fire's origin point. But there was no hint of apparent *odor*, which the accelerant invariably left. And Halliday could see that fire started in a highly inaccessible spot.

David Halliday: It would be quite difficult to apply a flame while the escalator wheels were moving past the track way.

The man in the blue boiler suit remained a unsolved mystery. It was clear he was a case of mistake identity. Two days after the blaze, the investigation team declared the arson theory was a dead end.

News anchor: Investigators still don't know what caused the fire but today they did rule out the possibility of arson. What no one can fully understand yet is the speed with which the small fire turned into intense fire ball.

If this was an accidental fire, it was even harder to explain how it suddenly turned into a *freakish* jet flame. And investigators still had to establish exactly how it started. With nothing else to go on, they went to the unburnt area under the escalator for clues. Here, they made a disturbing find, the wooden skirting boards all along the 42 meters' escalator appeared with *scorch* marks, not one or two, but 18 of them.

David Halliday: I concluded there was evidence of the old fires on the escalator.

8.7　Whether It was Caused by a Thrown Cigarette?

It was a shocking development. Something was triggering numerous small fires on the escalator. And one of them became the King's Cross blaze. There were scores of wooden escalators on London's tube system. Until the investigators traced the cause of the fires, it may only be a matter of time before Londoners faced another horrific inferno. Forensic scientists, investigating the London undergrounds' deadliest ever fire, have made a disturbing discovery. Evidences of the 18 old undetected fires under the escalator were the heart of the disaster. But the investigators were spotting something curious. Nearly all the scorch marks from the small fires were sided beneath the escalator's right hand side. It was a crucial lead. The rule of act on the tube was that passengers walk up the left hand side of the escalators and stand on the right. One thing easier to do while standing is lighting a cigarette. Could smokers carelessly discarded matches and accidentally started fires under the escalator? But the smoking was banned on the tube for over 2 years. It might be implicated in the old fires, but could it really be the cause of the recent disaster? David Shillito and the team checked with the station staff and made a discovery.

David Shillito: People smoking wouldn't be punished in station. The smoking ban was totally ineffective.

The ban was not strictly enforced. The prime suspect for the King's Cross fire was now a careless smoker discarding a still burning match. But something made this fire take hold, while all the others hardly burnt out. *Forensic* scientist David Halliday believed the fire started close to the escalator's running track. But all its component parts were made of metal. There was nothing here to fuel the fire. He examined an undamaged section of track, and found there was a thick coating of *lubricating grease*. He knew the grease alone was hard to ignite. But when he examined it more closely, he found it studded with discarded tickets, sweet wrapper, *fluff* from clothing, human hair and rat fur. It was clear that the running tracks haven't been probably cleaned since they were built from the started *The Second World War*. Could four decades of garbage created *inflammable mixture*? Back in the police forensic laboratory, Halliday prepared a simple test on the grease and garbage mix. He applied the *naked flame* to *replicate* the fact of the *burning match*. The dirty grease immediately caught fire.

David Halliday: I have found one thing at the scene that might be capable of being lighted by flame. And the test had confirmed that it was lighted by a flame.

The test revealed when inflammable *debris* burnt, it heated the grease and the mix to a point of ignites. It was a startling finding. It appeared that the *horrify* ing *fire risk lurked beneath the scores of wooden escalators on London*'s tube network. But igniting the dirty grease by the *Bunsen burner* in the laboratory, didn't conclusively prove the dropped matches could start a fire in the real world. David Shillito and the investigation team realized the only way to find out for sure was to try to recreate the chain of events at King's Cross station.

51 days after the disaster at 2:00 in the morning, the team gathered in the *eerie* setting of the discarded Piccadilly Line escalator. This original footage *captured* their *audacious* experiment. London's firefighters stood by to contain any blaze. Then on the undamaged lower half, one of the teams dropped a lighted match through the one centimeter gap between step and side wall.

"Three, Two, One, Start". The investigators held their breath. Would the flame *ignite* the dirt filled grease? They may have to drop dozens of matches before they found out the truth.

"Some fluff get fired". Incredibly, the very first match dropped ignited the grease and detritus mixes. After 1 minute and 55 seconds, the investigators could see the tiny glow through the gaps between the steps. And in just under 7 minutes, flames started to leak out through the wooden steps. The experiment was *working with a vengeance*. Shillito and the team faced a dangerous dilemma. They were desperate to see how the fire behaved. But the risk was that the fire might suddenly ran out of control just like it did in the disaster. After losing a colleague to the disaster, the firefighters were all aware of the risks.

David Shillito: The fire grew and it continued to grow, until the hose firefighter brigade told us to put it out.

8.8　Whether it was Caused by " Piston Effect" ?

It was a huge step forward. The investigators were now convinced that the King's Cross fire was started by a discarded match, igniting dirt covered grease. But to prevent to repeat the disaster, they still had to solve its most fundamental mystery. What transformed the small blaze into a rolling jet flame in a matter of seconds? Investigators were certain that the small amount of lubricating grease would not be capable to create such a freakish phenomenon. Because it burnt too slowly. The hunts for the missing factor *intensified*. Now investigators considered a theory being added by the news media. It suggested that the phenomenon unique to underground stations could be to blame. When a train approached the underground platform, it pushed a rush of air ahead of it. It was called the *piston effect*. It helped keep the tube system *ventilate*. Did the high speed rush of air act like a set of giant bellows on the fire, triggering the flame throwing effect described by the witnesses? Investigators *retrieved* an examining station log from the day of disaster. They wanted to analyze the movement of trains in and out, to build up the pattern of air movement up the escalator shaft. They worked out that in the 10 minutes before the inferno, trains departing underground platforms would pull fresh air down the escalator shaft. This would not affect any major effect on the fire. But they discovered that at 7:42, just 3 minutes before the fire erupted, everything changed. An eastbound Piccadilly Line train pulled in, seconds later a westbound train arrived, too. The combination of two trains arriving almost *simultaneously* had a dramatic impact on the *airflow*. It sent a huge blast of air up the escalator, which was the direction the fire traveled. It was a promising lead. To find out if it was the key to the disaster, investigators had to calculate if the airflow was traveling fast enough to have a significant effect on the fire, see Fig. 8-3. They measured the airflow produced by two trains, and calculated that the two trains arriving one after another would make the airflow up the escalator travel at 3.25 meters per second. That was only 12 kilometers per hour, not fast enough to trigger the jet flame. David Shillito and the team were not close explaining how insignificant escalator fire transformed into a rampaging killer.

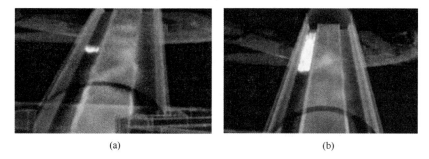

Fig. 8 – 3 Fire Simulation of Piston Effect

8.9 Find out the Reason of the Eruptive Fire

They ruled out terrorism, arson and the effect of airflow, and after exhausting search of the escalator shaft, they couldn't find nothing that might be acted like a super fuel on the blaze. The investigators desperately needed some fresh thinking. They turned to one of the most famous seat of learning——Oxford, and the budding science of computer modeling. Dr. Dienjons and Dr. Suzanne Simcox were pioneered to use the world fastest super computers to predict heat flows through buildings such as *nuclear reactors*. Investigators hoped their software might preview whether something from the *configuration* of King's Cross fire zone could have triggered the inferno. Simcox created precise computer model in the King's Cross underground station, and started a small virtual fire. As it started to spread, something odd happened. The fire started to lie down and clung to the surface of the escalator's steps. But basic physics dictated that it should burn straight up toward the ceiling. It was so *bizarre* that Simcox initially felt she had made an *elementary* mistake.

Suzanne Simcox: When we first showed the result, some people said did they set the gravity up side down? It was possible in computer modeling, since we set which the direction of the gravity is. We went to check again. And found we have got gravity the right way up.

There was no mistake, the fire continued to lie low on the escalator steps. Then within 30 seconds, it surged up the escalator at a high speed. Just like the jet fire eye witnesses saw at King's Cross. The scientists felt perplexed. It didn't make **sense**. It was an impossible fire. All they could tell the investigators was that it appeared something about the physical configuration of the escalator made the fire behave as it did. There was only one way to prove conclusively what it was. They must try to recreate the King's Cross fire for real.

8.10 Simulation Test Confirmed the " Trench Effect"

On July 1988, **nine months** after the disaster, 290 kilometers north from London, in the heart of Derby countryside, David Shillito and the team mounted an ambitious experiment. In the field renting from a local farmer, they built a model of Piccadilly Line escalator and the ticket hall, using all the correct materials. They rigged up a series of heat sensors and cameras. They were going to start a fire, on this time they would not intervene to put it out, Whatever it happened. A technician set light to the escalator in exactly the right spot. The investigation team and computer scientists watched anxiously. This could be the only chance to discover what caused the King's Cross fire. 6 minutes in, the blaze behave like a normal wood fire, with its flames standing up straight. At this rate, it could take up 60 minutes for the test fire to reach the ticket hall. As the seconds took by, it was beginning to look like the computer model got it wrong. Investigators were staring defeat in their face. If the experiment proved a dead end, the mystery of the King's Cross inferno may remain unsolved forever. Investigators were facing another dead end. Then 7 and a half minutes in, the fire dramatically changed. Rather than going straight up, the flame suddenly changed direction. Now they lay down and started to cling to the wooden steps of the escalator, just like they did in the computer model, as shown in Fig. 8 – 4.

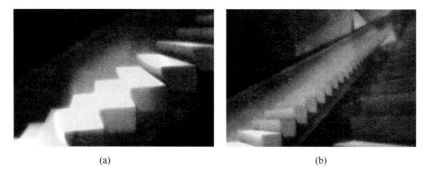

(a) (b)

Fig. 8 – 4 Simulation Test Confirmed the "Trench Effect"

Temperature sensors revealed that the wooden steps above the fire were now heating up very fast. Within 20 seconds, the investigators recorded temperature of 800 degree Celsius. Seconds later, the flame erupted unstoppably up the wooden escalator and burst into the model of ticket hall. They realized there was not one but two phenomena combined to create the King's Cross inferno. The 30 degree incline of the escalator had the effect of making the flames lie down on the steps. And the side of the escalator boxed the fire in, creating a channel or *trench* which concentrated the heat and flame and prevented them escaping. As the fire spread up the incline, it pushed before an invisible blanket of heat and gases, which superheated a 20 – meter stretch of wooden steps above. Around 10 seconds later, the superheated steps reached *critical temperature*, between 500 and 600 degree Celsius, and ignited. This created the devastating jet flame that surged so swiftly up the *escalator shaft*. It was the phenomenon that the scientists had never witnessed before. The team named the discovery as the *trench effect*, as shown in Fig. 8 – 5. It was the final piece of the puzzle. Investigators could now understand the chain of events that triggered King's Cross fire. What originally started the fire? What fueled its growth? And what triggered giant jet flame that devoured the ticket hall, and let 31 people sank from disaster.

Fig. 8-5 In-situ Test of the Eruptive Fire

8.11 Conclusions of Investigators

On Wednesday of November 18th, 1987, towards the end of the rush hour, a passenger riding of Piccadilly Line escalator dropped a match. By chance, the match fell through the gap in the wooden steps and was still burning on the rubbish clogged grease of escalator running track.

At 7:29 pm, 16 minutes to catastrophe, fire took hold beneath the escalator, commuter Philip Squire glimpsed the glow, he told the ticket collector, but when they looked from the top, the 30 degree angle hid it from eyes.

2 minutes to the disaster, fire fighters arrived and assessed the fire. They decided to be easy to contain. What they didn't know was that because of trench effect, gases from the fire were invisibly superheating the wooden steps.

20 seconds to go, as Brody and fellow passengers stepped into the ticket hall, the superheated stretch of escalator *spontaneously* ignited. Disaster struck, flames surged up the wooden steps traveling in 12 meters per second. A jet fire erupted into the ticket hall, with the heat it put of 2500 *megawatts*. That was as much energy as a jumbo jet used during taking off. 31 people caught up in the ticket hall inferno, died from burn and inhalation of toxic smoke. 11 months after the disaster, the investigation team finally delivered its report. It concluded the inferno's speed and scale were caused by the trench effect. It also ruled that all the wooden escalators be replaced with metal within 9 months. London Underground started *rigorously* enforcing the smoking ban straight after the fire. It also went on training its staff in the safest and quickest escape routes. Thanks to applying newly experimental work, fire teams around the world are now alert to the risk it poses by the blazes on *sloping surfaces*. Then you found knowledge will help save lives, should such inferno threaten again?

Words and Expressions

flamethrower	火焰喷射器，喷火器	tragedy	悲剧
commuter	定时往返的人，通勤族	tube	地铁，管道

surge 涌入
manhole 检修孔
dilapidated 破旧的
infrastructure 基础设施
smallholding 小农场
loyalty 忠诚
blaze 火焰
evacuation 疏散，撤退
firefighter 消防队员
horrible 可怕的
helmet 头盔
toxic 有毒的
iron 铁
budge 让步，动摇，微微移动
intense 强烈的
crawl 缓慢行进，爬行
gloom 昏暗，阴暗
inhalation 吸入，吸入物
paralytic 麻痹的，瘫痪的，中风的
victim 牺牲者，受害者
traumatic 令人痛苦的
wreckage 残骸，碎片
blowtorch 喷灯
chilling 令人恐惧的，吓人的
arson 纵火
accelerant 促进剂，加燃剂
arsonist 纵火犯
odor 气味
freakish 奇怪的
scorch 烧焦处，焦痕
forensic 法庭的，法院的
fluff 绒毛
grease 油膏，润滑油
replicate 复制
debris 碎片，残渣
eerie 可怕的
horrify 使震惊，使感到恐怖
capture 夺取，引起
audacious 大胆的

ignite 点燃
intensify （使）增强，加剧
ventilate 通风，使通风
retrieve 取回
simultaneous 同时的
airflow 气流
configuration 配置，构造
bizarre 奇怪的
elementary 基本的，初级的
trench 沟，渠
spontaneous 自然的，自发的
megawatt 兆瓦
rigorous 严格的，严密的
rush hour （上、下班的）高峰期
full swing 全面展开
subway system 地铁系统
computer simulation 计算机模拟
subterranean labyrinth passage
地下错综复杂的通道
soak up 吸收
descend into 向下行
tube system 管道系统
underground worker 地铁工作人员
emergency button 应急按钮
fire brigade 消防队
emergency exit 紧急出口
dozens of 很多
oxygen tank 氧气筒
underground structure 地下结构
track down 追寻，追查出
terror campaign 恐怖活动
steel work 钢结构，钢铁工程
criminal act 犯罪活动
lubricating grease 润滑脂
The Second World War
第二次世界大战
inflammable mixture 易燃混合物
naked flame 明火
Bunsen burner 煤气喷灯

burning match   燃烧的火柴
working with a vengeance   拼命工作
piston effect   活塞效应
nuclear reactor   核反应堆
critical temperature   临界温度
escalator shaft   电梯井
trench effect   沟槽效应
sloping surface   坡面

Translation Examples

[1] Now by rewinding the events back the whole day, and by going deeply into investigation, we can reveal what really happened in the King's Cross inferno. Advanced computer simulation will take us where no camera can't go, into the heart of the disaster zone.

现在我们将回溯当天的所有经过，并展开深入调查，揭露出国王十字车站大火的来龙去脉。先进的电脑模拟，将带我们来到摄影机看不到的地方，深入灾难现场的核心。

[2] All of the steel work and structure of the escalator was in fact generally intact. Where has distorted was clear to be distorted because of heat not the explosion.

所有钢结构和电梯结构，其实大致上完好无缺。扭曲的地方，一看就知道是因为热度而非爆炸造成的。

[3] When a train approached the underground platform, it pushed a rush of air ahead of it. It was called the piston effect. It helped to keep the tube system ventilate. Did the high speed rush of air act like a set of giant bellows on the fire, triggering the flame throwing effect described by the witnesses?

列车进入地下月台时，会推动前面的一阵气流，这叫作活塞效应。有助于维持地铁系统的空气流通，高速气流对火势的影响是否像一个巨大的风箱，触发了目击者描述的火焰喷射器效应？

[4] Around 10 seconds later, the superheated steps reached critical temperature, between 500 and 600 degree Celsius and ignited. This created the devastating jet flame that surged so swiftly up the escalator shaft.

过了10秒钟左右，过热的台阶达到临界温度，大约是摄氏500到600度，开始起火燃烧。因此形成毁灭性的喷发式火焰，快速地从电扶梯井往上冲。

Activities — Discussion, Speaking & Writing

Presentation

Group: 5 to 7 members

10 minutes per group (Each member should cover your part at least one or two minutes).

Clearly deliver your points of the following questions to audiences.

NEED practice (individually and together)!!

Gesture and eye contact.

Smile is always KEY!! Cover your nervousness!!

Questions for discussion and presentation

A disaster on this scale raises several crucial questions:

1. What triggered the blaze fire and what transformed the minor fire into a deadly jet flame in seconds?
2. Whether it was an arson or a terrorist attack? Why not?
3. Will the flame ignite the dirt filled grease? Why not?
4. Whether it was caused by thrown cigarette? Why?
5. Whether it was caused by "piston effect"? Why?
6. What is the reason of the eruptive fire?
7. What originally started fire?
8. What fueled its growth? And what triggered giant jet flame that devoured the ticket hall?
9. What is the "trench effect"?

Writing

Read **King's Cross fire**. Divide the following text into three parts. Write a short essay in English by groups which contains the following information.

英国伦敦地铁国王十字车站大火

1987年11月18日，英国伦敦地铁国王十字车站发生了一起31人死亡、大量人员受伤的特大火灾。这是世界地铁史上继1903年巴黎地铁发生死亡84人的大火后又一起罕见的地铁大灾难。1863年开通的伦敦地铁，是世界上最早的地铁，在1987年国工十字车站是伦敦地铁网络中最繁忙的车站，每天的人流量达25万余名，早晚高峰期间达10万余名。该车站是地铁皮卡迪利线和维多利亚线的共同车站。

人们开始怀疑那起大火是恐怖分子破坏造成的，但科学实验最后证明是点烟的火柴梗所致。这起大火过去30年了，但不少教训仍然值得我们吸取，关于火因调查的科学方法，也值得我们学习。

一、事故经过

当天晚上19时25分左右，一名乘客点烟后把火柴梗扔在皮卡迪利线4号自动扶梯上，火柴梗通过扶梯踏板与踢脚板的间隙落到积满油脂和可燃屑粒的自动扶梯运行钢轨上，引发火灾。

19时29分，一名乘客发现踏板下的小火，向售票员报警。1分钟后，另1名乘客启动了报警器。当时现场有2名警官，其中1名奔到地面上去通知伦敦消防总队，另1名警官决定经维多利亚线的自动扶梯穿过地铁的售票厅疏散车站下的乘客。

19点43分，伦敦消防总队和负责人到达现场。19点43分至45分之间，自动扶梯下发生轰燃，由于"沟槽效应"，火焰、黑烟喷入售票厅，造成包括消防队总指挥员在内的31人死亡。

二、惨痛的教训

1. 地铁站内设施落后，火灾隐患严重

起火的自动扶梯是1939年安装的，扶梯的踏板用有金属背面的胶合板制作，竖板用橡木制作，扶手用橡胶制作，附体踏板与踢脚板之间的防火条有30%不见了，踏板下面的运行轨道积满了油脂和可燃屑粒，从该自动扶梯使用到现在从来没有彻底清洗过。在20世纪的五、六十年代，有人提出为自动扶梯安装感烟探测器，但当局没有采用。

在发生火灾的时候，地铁售票厅中竖起一道隔开车站北部的临时木栅墙，这道墙堵住了通往第4部楼梯的入口，挡住了室内消火栓和伦敦消防总队的一只灭火计划箱。

2. 地铁公司消防意识淡薄

国王十字车站使用的这类自动扶梯特别容易发生火灾，据统计，在1939年至1944年间就发生过77起，1944年11月24日在派丁登的日克罗线自动扶梯发生了特别严重的火灾，使内部装置完全被毁坏。但由于以前的火灾没有人员死亡，伦敦地铁当局对此掉以轻心，认为有乘客和职工起探测器的作用，发生火灾时有足够的时间安全疏散乘客。

3. 缺乏必要的消防措施

在这次大火发生之前，伦敦地铁公司缺乏必要的消防措施，员工缺乏必要的消防训练，车站没有疏散计划，通道设备差，全车站没有人负责全面的安全监督管理。

三、给我们的教训启示

三十年过去了，我国三十多个城市都有了地铁，截至2018年10月，中国已开通的城市地铁（以首条轨道交通开通时间排序）有36个，总里程已居世界第一。1993年开通的上海轨道交通为世界上规模最大、线路最长的地铁系统。中国大陆首条地铁系统是北京地铁，建于1965年，竣工于1969年，试运营于1971年1月。

然而，当年伦敦地铁国王十字车站火灾的经验教训对我国地铁的消防安全工作仍然有值得借鉴的地方。木质的自动扶梯在我国的地铁里从来未使用过，如今我国地铁的消防设施和管理也远比当年伦敦的先进，情况有很大的不同，但是，地铁车站一旦发生火灾、浓烟积聚、蔓延迅速、疏散困难和扑救困难的特点依然没有改变，因此我们应汲取伦敦地铁火灾的教训。

（1）轨道交通服务，不能容忍"火灾危险在一定程度上可接受"的观念。在1987年前的一百多年里，伦敦地铁发生过几十起火灾，国王十字车站也发生过数起小火，都没有人员死亡。因此没有引起伦敦地铁公司和国王十字车站的管理层足够的重视，导致对许多火灾隐患熟视无睹，留下了火灾的温床。消防安全管理制度必须令行禁止。伦敦地铁里有禁止吸烟的规章制度，但没严格执行，结果是一名吸烟乘客扔下的火柴梗引发了火灾，这本来是可以避免的。

（2）员工良好的消防培训，能及时化险为夷。国王十字车站那天当班员工中，只有4名受过消防培训。自动扶梯起火时，发现的员工没有受过消防培训，不知道当时应该开启水喷雾装置，而是去远处取二氧化碳灭火器，等他赶回来时，火势已大，灭火器也无济于事。

（3）高效的应急管理体制是对付重大灾难的有力措施。缺乏高效的应急管理体制是

国王十字车站特大火灾导致31人死亡的重要原因之一。发生火灾时，调度室里没人；售票厅里的乘客没及时疏散；十几分钟后，还有列车在该站停车下客。没有垂直的安全管理系统，是伦敦地铁公司应急管理上的缺陷。无论在地铁公司还是车站，安全工作分别属于各部门管理，缺乏完整的顶层设计，发生火灾时各行其是，现场混乱是自然的了。

四、值得学习的经验

国王十字车站特大火灾留下了必须汲取的教训和值得学习的经验。火灾发生后，英国运输部负责事故调查。调查的主要目的是查清三个核心问题，即怎么起火，为什么大火发生如此突然和为什么会造成31人死亡。在调查的过程中，他们组织了专家组，做了大量实地调查，召开听证会，对伦敦地铁火灾史、管理体制和地铁管理人员对消防安全的认识等许多方面进行调研，还根据火灾现场的勘察需要，做了科学试验，弄清火灾现场形成轰燃的条件。通过历时一年的调查，最后形成《伦敦地铁国王十字车站调查报告》，报告介绍的安全观念、火灾教训、火因鉴定程序和方法、整改建议等对地铁的消防安全工作很有参考价值。

另外，有关部门事后还对火灾隐患进行了全面的整改，对消防工作的薄弱环节做了加强，使英国地铁的消防安全水平有显著的提高。

Unit 9

Leap through Time-Earthquake

Teaching Guidance

Imagine you are walking along a street in Japan. Suddenly, there is great *roar* and the ground starts to shake violently beneath your feet. You are in great danger! For the next few minutes, many things happen around you, following on one after the other with frightening speed…

The story told in this unit is like a journey. It is not a journey you can make by plane, car or ship. In fact, you don't have to go anywhere at all. You are about to travel through time. With each turn of the text, the time moves forward a few seconds, hours or even years. Each new time—each stop on your journey is like a chapter in the story. The first violent shaking, the toppling of buildings, the cracks opening up in the ground, the *collapse* of bridges, the *landslide*, the vast waves engulfing the *shore*, the fires sweeping through the devastated city, the *rescue of survivors* from *beneath the rubble*, the building of a museum years later to record the events of those fateful moments——all tell the story of the great earthquake.

9.1 About 3000 Years Ago

It is late afternoon on the island of Honshu, Japan. A small band of hunters are returning to their village along the coast path. As they start to climb down through the woods, they are startled by a sudden loud, thundering noise.

Looking up, they see the *cliff* on the opposite side of the bay collapse into the sea. Out in the water, they see waves start to rear up several meters high and crash on to the shore. Their village, built on the flat land close to the water's edge, will certainly be flooded. It is a horrifying sight.

The men know that an earthquake has struck. This is not an unusual event—small ground *tremors* lasting a few moments occur almost weekly, but no damage or injury is caused. This quake is different. The men have never seen so much of the cliff collapse before, nor witnessed such *colossal* waves…

WHAT IS AN EARTHQUAKE?

The outer layer of the Earth is made up of a number of giant slabs, called *tectonic plates*. These are always on the move. In some places, the edges are moving apart. In others they are colliding, with one plate sliding underneath the other. In still others, one plate edge slides alongside another plate. This movement is very slow——about 1cm a year but the pressure is enormous. When plate edges grind against one another, they send out shock waves through the ground. We feel these as the *vibrations* which are earthquakes.

Most earthquakes are small tremors that do no damage. But when sliding plates lock together, pressure builds up in the rocks underground. Eventually the pressure becomes too much for the rock to withstand. It snaps, causing a major earthquake.

The place where the rock breaks is called the focus of the earthquake. The point above it on the Earth's surface is termed as the epicenter.

Japan suffers from many earthquakes because it is situated near plate boundaries, The Pacific plate is sliding down beneath the Eurasian plate.

9.2 A Hundred Years Ago

The village has grown over the years and has now become a prosperous city with a busy sea port. The streets are crowded with carts and rickshaws. Street sellers offer all kinds of things for sale, including food, clothing, clogs, lanterns and toys. Amid all the noise and bustle, a funeral procession slowly winds its way through the crowds. In one of the buildings, a tea *ceremony* is being held.

Over the years, the city has suffered many earthquakes. Most have been far too weak for anybody to notice, although some stronger quakes have caused some minor damage. The city's buildings have been constructed in a particular way, which makes them very resistant to earthquake shaking. Walls are made of wood and, between rooms inside the houses, with paper. The wooden beams holding up the thatch roofs are tied with rice straw ropes.

9.3 A Few Years Ago

The city is a modern *metropolis*. Many of its traditional buildings have been demolished and replaced by *multistory blocks*, built of steel and glass. This morning the weather is unusually hot and humid. The *fog* rolling in from the sea has been slow to lift. Cars and buses take people to work along the city streets as usual, but today there is a strange feeling in the air. Animals, including sea birds and people's pet dogs, are restless and jumpy. Some fishermen stand about and chat on the pavement.

Unusually, they have caught no fish at all that morning. Many report having seen strange lights in the sea fog earlier. One old man, who remembers witnessing very similar events many years ago, is in no doubt as to what all this means. He tells whoever will listen: there will be an earthquake soon.

9.4 Later That Day

The old man is right. An earthquake strikes that very afternoon. With a deafening roar, the whole city starts to shake violently. Within seconds, large *concrete slabs* start to fall off the outsides of buildings and crash to the ground. *Glass panes* fall and shatter on the pavements below. In some buildings, the walls fall outwards, causing the concrete floors to collapse on top of one another like a pack of cards. Sirens wail amid the din of falling rubble.

People dash out into the street, screaming in panic. As they do so, great cracks in the road's surface start to open up. Cars *swerve* to avoid the cracks and screech to a halt. Some collide with each other, or smash into buildings. Water from burst pipes and underground drains pours out on to the streets. Electricity cables spark and crackle. Everywhere, there is *chaos*.

9.5 Seconds Later

Inside an apartment, a family just sitting down to a meal find themselves being hurled around their kitchen. The rocking and juddering cause plates to tip out of cupboards, pots and pans to crash to the floor and even the furniture and electrical appliances to slide around the room, and cracks appear in the ceiling and lumps of plaster start to crash down amid clouds of dust.

The children scream. Their mother shouts to them to get under the table. Everyone is terrified. It feels like a sickening roller-coaster ride where everything is breaking up around them.

UNSTABLE GROUND

When shock waves from an earthquake pass through *solid rock*, it is completely unaffected. But shock waves passing through *sediments* such as *moist sand* or gravel cause them to become almost like a liquid. This is called *liquefaction*. Buildings sink into the liquefied sediments and *topple over*. Mexico City was built on old lake sediments, so many buildings were badly damaged when an earthquake struck the region in 1985. The districts of San Francisco that were damaged in the 1989 quake had been built on top of debris from the 1906 earthquake that had been dumped in the waterfront area.

9.6 At the Same Time

Just outside the city, farmers in the rice fields are flung to the ground by the force of the earthquake. To the alarm of the terrified onlookers, the land itself rises and falls like the surface of the ocean as waves surge across it. In the fields, *fountains* of mud spurt into the air like miniature *volcanoes*. Cracks appear in the field and a nearby road buckles, throwing cars off the edge.

With a deafening crack, a railway bridge suddenly give way as its concrete supports collapse. The tracks start to slump. The train driver slams on his brakes in a desperate attempt to stop his train from plunging to the ground.

FAULTS

Rocks can bend and fold without breaking. Sometimes, when subjected to great pressure, they may suddenly break. The crack in the rocks where this sudden movement takes place is called a *fault*. As the pressure continues, rocks may move past one another along the same fault. During an earthquake, portions of land may be raised up or slip down along fault lines.

9.7　A Few Minutes Later

After 20 seconds, the quake is over. But the violent shuddering of the ground has not only caused buildings to collapse. Up on a hillside overlooking the city, the rocks and soil just beneath the surface have been made unstable by recent heavy rain. Now, triggered by the quake, the slope itself gives way. In a gigantic landslide, tonnes of boulders and soil, together with trees and shrubs, start to surge downhill. Houses, cars and anything else in the way of the landslide are carried along with it, so adding to the slide. Alerted by the thunderous noise of falling rocks, people dash from their houses and flee for their lives…

9.8　Twenty Minutes Later

During the quake, some people did not have time to escape before the buildings they were in collapsed. *Miraculously*, a few strong beams held up parts of the building in this street and saved people from being totally crushed. But they are now buried beneath piles of rubble. They have only dusty air to breathe and there is a risk that an *aftershock* (a lesser tremor that takes place after the main quake) will destroy their *fragile shelter*. They cry out, hoping someone will rescue them quickly.

Soon, a crowd of helpers, including firemen, emergency medical teams and some brave people who did manage to escape the falling buildings, rush to the scene. Listening out for shouts of help, they lift away the rubble and cut through wood and twisted metal.

Eventually a loud cheer goes up as the first survivors are hauled to safety. They are carefully stretchered over the rubble to a waiting ambulance.

Just then, an aftershock causes the mains of the building to collapse. The rescuers were just in time!

On the sea front, a group of people suddenly notice a massive wave out to sea, getting bigger all the time and rapidly moving towards them.

"*Tsunami*!" They yell, and everybody runs for their lives.

Soon, a wall of water with some 30 metres high *looms into view*. Ships and boats of all sizes are picked up and, as the giant wave surges across the docks, are *hurled on to the shore*. Many more tsunamis follow, one after the other, over the next few hours. A hotel building on the seafront escapes destruction by allowing the *torrents of water* to pass unhindered through its lower floors.

TSUNAMIS

Tsunami is a Japanese word meaning "a wave breaking into a harbor". Most tsunamis are caused by earthquakes on the sea bed. Some, triggered by quakes in one part of the world, do great damage to coastal areas in another part a long way away. The great Chilean earthquake of May

1960 created tsunamis that devastated Japan 16000km away.

9.9 Several Hours Later

Night has fallen, but the effects of the earthquake are far from over. Fire has broken out in the city. Sparked by fallen power lines and fed by gas escaping from broken gas pipes, the flames quickly fan out among the stricken city's buildings. Moreover, broken water pipes make it very difficult for the firemen to put out the fires. Here they are using a pump to bring in water from other parts of the city where pipes are still intact. Elsewhere, they will need to blow up some buildings to stop the fire from spreading.

People who have been rescued from fallen buildings sit together in the street as the teams of firefighters attempt to save their homes from even *further destruction*. Meanwhile, rescue workers continue their search for survivors. They still have hopes that more people can be brought out from under the rubble. They will work on through the night.

9.10 The Next Morning

Dawn breaks on a scene of utter *devastation*. A reporter flies over the city in a helicopter, relaying to his shocked radio listeners what he can see below. Many buildings are damaged or totally destroyed. Large sections of the elevated roadway have collapsed. Fire and flooding have ravaged much of the city.

But the story is not wholly *gloomy*. Many buildings, specially built to withstand earthquakes, have remained undamaged. They include a number of old buildings, with their wooden or paper walls and *thatched roofs*. And, despite the great ferocity of the quake, most of the city's inhabitants escaped harm. They have spent the night sleeping in tents. For the next few weeks, they will be without electricity and running water, but at least they are alive.

9.11 Today, a Few Years Later

A party of school children are visiting a museum. It was built recently as a record of the terrible events of a few years ago, in which the city suffered its worst ever earthquake.

The exhibits tell people all about earthquakes. There are models showing the Earth's plates and what a fault in the rocks looks like. The visitors watch a video of what happened during the quake in their city, and *gaze* at a model of the immense damage it caused.

There are also exhibits that show how traditional houses were built to withstand earthquakes. A *seismograph* records small tremors as they actually happen in the city today.

Children go on an earthquake *simulator*. The floor shakes around, making it very difficult for them to stand upright. The children have a lot of fun falling about, but their parents remember that when the real earthquake struck, the shaking was so violent that they feared for their lives.

Unit 9 Leap through Time-Earthquake

Words and Expressions

roar 轰鸣声	dawn 黎明
collapse 崩塌	devastation 毁灭
landslide 滑坡	gloomy 阴暗的
shore 海岸	gaze 凝视
cliff 悬崖	seismograph 地震仪
tremor 震颤	simulator 模拟器
colossal 巨大的	rescue of survivor 抢救幸存者
vibration 振动	beneath the rubble 废墟下面
epicenter 震中	tectonic plate 构造板块
ceremony 典礼,仪式	wooden beam 木梁
metropolis 大都市	multistory block 多层建筑
humid 潮湿的	concrete slab 混凝土板
fog 雾	glass pane 玻璃板
swerve 突然转弯,突然改变方向	solid rock 坚固的岩石
chaos 混乱	moist sand 潮湿的沙子
sediment 沉积,沉淀	topple over 倾倒
liquefaction 液化	fragile shelter 脆弱的避难所
fountain 喷泉	loom into view 隐约可见
volcano 火山	hurl on to the shore 猛冲到岸边
fault 断层	torrent of water 急流
miraculous 奇迹般的	further destruction 进一步破坏
aftershock 余震	thatched roof 茅草屋顶
tsunami 海啸	shock wave 冲击波

Further Reading

1. 143000 people died in the 1923 Tokyo quake.

2. Thousands of earthquakes occur every year, some of them are major ones. But only where cities are affected will be great loss of life.

3. Seismographs measure the size of shock waves caused by an earthquake. Vibrations are recorded by a pen on a paper as a trace rolled over a drum (see Fig. 9 – 1).

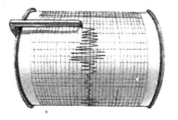

Fig. 9 – 1 Seismographs

4. The Chinese invented the earthquake detector (see Fig. 9 – 2) in AD 132. When shaken, a rod inside it swings and opens one of the dragons'

mouths, releasing a ball into a toad's mouth with aloud "dong". It records the direction of the quake.

5. When rocks snap, two kinds of shock waves are released from the earth quake's focus. Primary (P) waves (see Fig. 9 – 3) squeeze and stretch the rocks. Secondary (S) waves (see Fig. 9 – 4) shake them up and down and from side to side.

6. Globes in Fig. 9 – 5 show where earthquakes occur around the world. Dashed lines in globes show the plate boundaries. Most take place at or near plate boundaries. The biggest are where plate edges slide alongside, or where one slides under another.

Fig. 9 – 2 Earthquake Detector

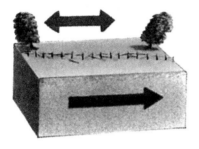

Fig. 9 – 3 Primary (P) Waves   Fig. 9 – 4 Secondary (S) Waves

(a)   (b)

Fig. 9 – 5 Globes Show Where Earthquakes Occur around the World

7. The San Andreas Fault stretches more than 1200km along the coast of California. It forms the boundary between two plates that are sliding past one another in opposite directions. The jerky movements result in constant, tiny shocks. Occasionally pressure builds up over the years and is released in a massive quake, such as the 1906 event in San Francisco.

8. When the epicenter of an earthquake is on the sea bed, a large submarine landslide can happen (see Fig. 9 – 6). This produces a series of fast moving waves, travelling at around 800km per hour, called tsunamis (also sometimes known as tidal waves). In deep water, they are small, but as they approach the shallow coastal waters, they slow down and build up in height. Some tsunamis are tens of metres high as they crash onto the shore (see Fig. 9 – 7). The waves may last for several hours.

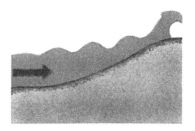

Fig. 9 – 6 Epicenter of an earthquake on the sea bed

Fig. 9 – 7 Some tsunamis with tens of metres high

9. The Mercalli Scale records the intensity of damage caused by an earthquake. The following chart shows the descriptions of different levels in Mercalli Scale.

Intensity of The Mercalli Scale Table 9 – 1

Intensity	Descriptions
I ~ II	Felt slightly
III - IV	Vibration as if by passing lorry (Fig. 9 – 8 shows the IV intensity)
V	Buildings tremble, vases fall, trees shake (as in Fig. 9 – 9)
VI	Bells ring, plaster cracks, people shake
VII	Tiles loose, old walls fall, chimneys crack
VIII	Damage to buildings
IX	Ground cracks, buildings collapse
X	Landslides, bridges damaged, rails bent
XI	Dams wrecked
XII	Total devastation

Fig. 9 – 8 IV damage

Fig. 9 – 9 V damage

10. Building in San Francisco is specially designed to withstand even the most severe earthquakes. It is shaped like a slender pyramid and, thanks to special steel supports, its top is strong enough to sway 12m, still remaining intact. In 1989, an earthquake shook California. Its epicenter was 100km south of San Francisco. Although quite severe, the quake damaged few buildings in San Francisco. Most had been built to resist earthquake shaking.

There are several ways in which buildings can be constructed to withstand severe quakes.

Walls are anchored to concrete foundations reinforced with steel rods.

Steel brackets anchor brick chimneys to the roof. Metal chimneys are lighter and safer.

Boilers are held in place by straps bolted to the wall, preventing gas pipes from breaking.

Steel connectors reinforce the joins between the wooden beams and joists supporting the floors and ceilings.

Activities—Discussion, Speaking & Writing

Presentation

Group: 5 to 7 members

10 minutes per group (Each member should cover your part at least one or two minutes).

Clearly deliver your points of the following questions to audiences.

NEED practice (individually and together)!!

Gesture and eye contact.

Smile is always KEY!! Cover your nervousness!!

Questions for discussion and presentation

1. What are basic requirments of seismic design according to the Code for Seismic Design of Buildings?

2. What should we do to save ourselves in case of earthquake disaster?

Writing

Search for earthquakes in the American Engineering Index and the American Scientific Citation Index. Then write a report independently on "What can we learn from unthinkable earthquake?" The following table shows examples of some enormous earthquakes around the world.

Examples of Enormous Earthquakes Table 9 – 2

	Alaska Earthquake, USA	San Francisco Earthquake	Kobe Earthquake, Japan	The Kanto Earthquake, Japan	Chile earthquake	The Indonesian tsunami	Mexico Earthquake	Tangshan Earthquake	Wenchuan Earthquake
Time and Site									
Magnitude									
Secondary disasters									
Disaster Profile									
Present Development									
Group Members									

Unit 10

Green Buildings

Teaching Guidance

The earth is the cradle of life, the common homeland of mankind.
Protecting the planet is our common responsibility.

Robert Watson, NRDC, Founding Chairman, LEED Steering Committee Director,
U. S. Green Building Council
美国绿色建筑协会缔造人,LEED 标准委员会主席,Robert Waston

NRDC: Founded in 1970, the association protects the planet through law, science, policy and human power. It has four offices, 250 professionals and more than 1 million members in the United States.

美国自然资源保护委员会成立于1970年,通过法律、科学、政策和人的力量保护地球,总共有4个办事机构,250名专业人员,全美有超过100万名会员。

10.1 Environmental Impact of Buildings

The main environmental impacts of buildings are the reduction of habitats and species due to land consumption, air pollution due to energy consumption, water pollution due to water consumption, resource loss due to material consumption, and indoor environmental quality closely related to human health.

According to statistics, buildings consume 70% of the electricity, 40% ~45% of the primary energy, 40% ~45% of the greenhouse gases, 35% ~40% of the municipal solid waste, and 80% of the drinking water in the United States. In China, 700 thousand hectares of land are developed each year, and 3 million acres of farmland are occupied by development projects.

建筑对环境的影响主要有土地消耗导致栖息地和物种的减少、能源消耗所致的空气污染、水消耗带来的水污染、材料消耗导致的资源损失还有与人类健康息息相关的室内环境质量。

据统计,建筑每年消耗全美70%的电能,使用全美40%~45%以上的一次能源,排放全美40%~45%的温室气体,建筑垃圾占城市固体垃圾的35%~40%,消耗美国80%的饮用水。而在中国,每年开发70万公顷土地,300万亩农田被开发项目占用。

10.2 What is "Green" Design?

Green design refers to the noticeable reduction or elimination of the negative impacts of buildings on the environment and users in architectural design and construction in five aspects, namely, the sustainability of site planning, safeguarding water and water efficiency, energy efficiency and renewable energy, conservation of materials and resources, and indoor environmental quality.

Sustainable development mainly includes three conditions: economy, social equity and environment. Among them, the economic conditions are mainly local economy, local materials and workforce, global economy; the social equity conditions are mainly the government, social justice and the historical continuity of some sacred places, material resources, human impact, as well as spiritual and physical health; the environmental conditions are mainly divided into natural system and man-made system, natural systems are sun, wind, water systems, animal and plant habitat, the man-made systems are human and manufactured waste, agricultural resources, transportation, energy systems, resource extraction impact.

New construction is the best time to capture the benefits of green building and transform the market. If opportunity is missed initially, waste and damage will last for many years. At the same time, the marginal cost is the lowest. Seize the opportunity of integration can result in additional benefits and eliminate extra costs.

绿色设计是指建筑设计和施工应在五个方面明显降低或消除建筑对环境和用户的负面影响，它们分别是场址规划的可持续性、保护水和提高水的使用效率、节能和可再生能源的开发利用、材料与资源节约和室内环境质量。

可持续发展主要有经济、社会公平和环境等三个条件。其中经济条件主要是地方经济、地方物资与劳动力、全球经济；社会公平条件主要是政府、社会正义和一些神圣地方的历史延续性，资源的来源和人类的影响以及身心健康；环境条件主要分为自然系统和人造系统，其中自然系统是太阳、风、水系统、动植物栖息地，人造系统是人类与制造废物、农业资源、交通、能源系统、资源开采的影响。

新建筑是抓住绿色建筑效益和市场转型的最好时机。如果失掉最初的机会，浪费和破坏将会持续很多年。而此时的边际成本最低，抓住整合的机会可以带来额外效益并消除额外成本。

10.3 Green Building Assessment Systems

The assessment systems of each country are as follows: United States, LEED™; Britain, BREEAM; Japan, CASBEE; Canada, GBTool; France, ESCALE; Norway, Eco Profile; Germany, LND; Australia, NABERS and in China the system is "Evaluation System for Green Olympic Buildings" (GBCAS).

The key attributes of successful construction standards are technically sound, green achievement, feasibility, market acceptability and program support.

Modern green building movement started with energy crises of the 1970s, which began with energy efficiency standards and codes, followed by indoor air quality, material issues, site and water, the first green buildings in U.S. dated back to 5,000 years ago, the earliest green building in America was Mesa Verde, it is a form of burrowing, living in caves.

U.S. Green Building Council, a national non-profit organization based in Washington, DC, was founded in 1993 with the goal of integrating the construction industry, leading the market transformation, and educating owners and practitioners. It is made up of diverse institutional members, developed and administrated by the LEED™ Green Building Rating System.

每个国家的评估体系如下：美国，LEED™；英国，BREEAM；日本，CASBEE；加拿大，GBTool；法国，ESCALE；挪威，Eco Profile；德国，LND；澳大利亚，NABERS，在中国，该系统是"绿色奥运建筑评价体系"（GBCAS）。

成功的建筑标准的关键属性是技术合理、绿色成果、可行性、市场可接受性和支持方案。

现代绿色建筑运动始于20世纪70年代的能源危机，从节能标准和规范开始，随后出现室内空气质量、材料、场址和节水问题，美国最早的绿色建筑可追溯到5000年以前，美洲最早的绿色建筑是梅萨沃德印第安遗址，是一种穴居的形式。

美国绿色建筑协会是设于华盛顿特区的非营利性机构，成立于1993年，目标是整合建筑业、领导市场转型、培训业主和从业人员。由多样性的机构会员组成，由"绿色建筑评估认证标准体系"LEED™开发管理。

10.4　Leadership in Energy & Environmental Design（LEED）

LEED™ is a leading-edge system for designing, constructing, and certifying the world's greenest buildings and was created to define "green" by providing a standard for measurement, preventing "greenwashing" (false or exaggerated claims), used as a design guideline and promoting whole-building's integrated design processes.

LEED™ point distribution includes the following items: Indoor environmental quality accounts for 23%, sustainable sites accounts for 22%, materials & resources accounts for 20%, water efficiency accounts for 8%, energy & atmosphere accounts for 27%.

Among them, sustainable sites includes erosion & sediment control, such as sediment & erosion control fencing, site selection, urban redevelopment, brownfield redevelopment, alternative transportation, reduced site disturbance, stormwater management, landscape & exterior design to reduce heat islands, like using green roofs in Gap corporate headquarters in San Bruno, CA, as well as light pollution reduction. There are two real cases of efficient site selection——PNC Bank in Pittsburgh, PA and Kandalama in Dambulla, Sri Lanka.

LEED™是世界上最绿色的建筑设计、施工与认证的先导体系，而创建这个体系是为了通过提供一个衡量标准来定义绿色、防止错误地或夸张地滥用"绿色"、作为设计指导和推动整座建筑的整合的设计过程。

LEED™评分标准包括：室内空气质量占23%、可持续性占22%、材料与资源占20%、节水占8%、能源和大气占27%。

其中，可持续性场地包括侵蚀与泥沙控制，如防沙栅栏、场地选择、城市再开发、被污染的土地再开发、可替代交通、减少对场址的影响、雨水管理、减少热岛效应的景观和外部设计，比如位于圣布鲁诺的GAP公司总部使用的绿色屋顶，还有减少光污染。这里有两个高校选址的实际案例，位于匹兹堡的PNC银行和位于丹布拉的坝达那玛饭店。

Secondly, energy & atmosphere includes the following items: Fundamental building commissioning, accordance with ASHRAE/IES 90.1-1999, alternative or renewable energy, elimination of CFCs, an additional commissioning, elimination of HCFCs and Halons. After that, the measurement and verification are taken, followed by signing green power contract. There are many actual cases, such as efficient lighting, "Super Windows", integrating natural and electric light, underfloor air, building controls and photovoltaic roof.

And then, the factors of Water Efficiency are innovative wastewater technologies and water use reduction, such as xeriscape and drip irrigation.

What's more, materials and resources include the following factors: Storage & collection of recyclables, building reuse, construction waste management, resource reuse, use of materials with recycled content and regional materials; we can also use rapidly renewable materials and certified wood like wheatboard and bamboo.

The last point is Indoor air quality(IAQ), this item means that in order to meet ASHRAE 62-1999, we should control environmental tobacco smoke and monitor CO_2 to ensure effective ventilation. Setting up construction IAQ management plan, like using low-emitting materials, indoor chemical and pollutant source control, is to ensure controllability of systems and thermal comfort.

其次，能源和大气包括以下几个项目：建筑的基本试运行，符合 ASHRAE/IES90.1-1999，使用替代或可再生能源，消除氟利昂，其他的试运行以及消除氯化烃和 HALON。之后，进行测量及检验，并签订绿色电力合同。有很多实际案例，如高效照明，"超级窗户"，整合自然光和灯光照明，地板下送风，楼宇自控，屋顶光伏发电。

水效率的影响因素是创新的污水处理技术和减少用水量，如耐旱地形和滴灌。

此外，材料和资源包括以下几个因素：可回收物的储存和收集、建筑再利用、建筑垃圾的管理、资源的再利用、再生材料的使用和本地材料的使用，我们还可以使用快速再生材料和认证的木材，如麦秸板和竹子。

最后一点是室内环境质量，该项目意味着为了满足 ASHRAE 62-1999 要求，我们应该控制室内吸烟和监测二氧化碳以确保有效的通风。建立建筑室内空气质量管理规划，如使用低排放材料，室内化学和污染物源控制，确保系统的可控性和热舒适性。

The technical overview of LEED™ are whole-building approaches to encourage and guide a collaborative, integrated design and construction process, optimize environmental and economic factors. Four levels of certification about LEED™ are LEED Certified, Silver Level, Gold Level and Platinum Level. LEED™ certification process is a three-step process: Step 1 is project registration, which includes welcome packet and on-line project listing; Step 2 is technical support, namely credit rulings, and Step 3 is building certification, upon documentation submittal reviewed by USGBC. And the benefits of certification is that the buildings can get independent recognition of quality buildings and environmental stewardship. The recognition is from a third party validation of achievement, the LEED™ Certification plaque can also be mounted on buildings as an official certification, which can receive marketing exposure through USGBC Website, case studies, media announcements.

LEED™的技术概要是整体的建筑理念鼓励并引导合作、整合的设计和施工过程、优化环境与经济因素。LEED™的4个认证等级分别是 LEED 认证级，银级，金级和白金级。LEED™认证程序分三步，第一步是工程注册，包括欢迎信息和在线工程一览表；第二步是技术支持，也就是指标认定；第三步是建筑认证，其中的编制文件由美国绿色建筑协会评审。LEED™认证的好处是：可以获得高品质建筑与环境管理的认同。这是第三方成就认证，还可以在建筑上嵌贴LEED认证标识作为一种官方认证，可以通过美国绿色建筑协会网站、媒体、案例研究向市场公布。

10.5 Non-Economic Benefits of Green Building

Non-Economic Benefits of Green Building includes environmental benefits, such as reducing the impacts of natural resource consumption; health and safety benefits, such as enhancing occupant comfort and health; community benefits, like minimizing strain on local infrastructures and improving quality of life. In addition, the economic benefits are competitive first costs and reduced operating costs, competitive first costs mean that integrated design allows high benefit at low cost by achieving synergies between disciplines and between technologies; reduced operating costs can be explained by cases that in the U.S., green buildings cost $ 0.50 ~ $ 0.60 per square foot to operate compared with $ 1 ~ $ 2 for a standard building.

绿色建筑的非经济效益，第一个是环境效益，如降低自然资源消耗带来的影响；第二个是健康与安全效益，如提高居住舒适性与健康性；第三个是社区效益，例如减轻对地方基础建设的压力，改善生活质量。而另一方面，绿色建筑的经济效益是有竞争的初投资和降低运行费用，第一个是通过增强建筑各专业和技术的协调，整合的设计可以较低的投入取得较高收益；第二个的实际案例是美国的常规建筑运行费用为每平方英尺 $ 1 ~ $ 2，而绿色建筑的运行费用降至每平方英尺 $ 0.50 ~ $ 0.60。

10.6 Green Buildings in China

Agenda 21 Project is a green building in China, The building has 8 storeys, with a total construction area of 13000 square meters, and the building are used for government offices, which will house ACCA21 and NSCTSD offices. The second floor will serve as Technology Demonstration and Training Center. It is a Chinese and American design collaboration.

21世纪议程项目是一个中国的绿色建筑。该建筑总共8层，总建筑面积13000平方米，建筑用途是政府办公楼。中国21世纪议程管理中心和中国科学技术促进发展研究中心将在此办公。其第二层将会作为节能技术展示厅及培训中心。该建筑是中美合作设计。

10.7 Green Building Features

The features of green buildings are mainly in the following aspects: Efficient site location, which means infill development and near mass transit, ensuring lots of bicycle parking and reducing auto parking; Roof garden, which means rainwater capture and urban heat island mitigation; Replanting trees, by which we can landscape with native vegetation; Water efficiency, which means use of waterless urinals and dual flush toilets; Green finish materials, such as recycled content and low VOCs; Advanced air filtration, such as Dynamic Air Cleaners. All of above is to get to LEED Certified level.

绿色建筑的特征主要有以下几个方面：位置效率，意味着成熟社区内开发和临近公共交通设施，保证充足的自行车泊位和减少汽车泊位；屋顶花园，是指雨水收集和降低城市热岛效应；树木的移栽，是指采用本地植物的景观布置；节水，使用无水小便器和双冲厕所；绿色装修材料，使用再生成分地毯和低挥发性有机物；先进的空气过滤，使用动态空气过滤。以达到 LEED 认证标准。

Further Reading

Reading Material

Introduction to BIM(Building Information Modeling)

1. *What is BIM*

Building information modeling (BIM) is not a specific software program. It is a *streamlined* process that allows us to make better decisions about project design based on reliable information analysis. BIM is an approach to the entire project life cycle, including design, construction, and *facilities management*. The BIM process supports the ability to *coordinate*, update, and share design data with team members *across disciplines*. The 3D process is aimed at achieving savings through collaboration and *visualization* of building *components* into an early design process that will *dictate* changes and *modifications* to the actual construction process. It helps engineers better predict a project's performance to increase safety, *constructability*, and *sustainability* before it is built, thus facilitating better decision and making more economic project delivery.

2. *BIM Throughout the Project Life-cycle*

Use of BIM goes beyond the planning and design phase of the project, extending throughout the building life cycle, supporting processes including *cost management*, *construction management*, *project management* and *facility operation*.

(1) *Management of building information models*

Building information models *span* the whole concept of occupation time-span. To ensure efficient management of information processes throughout this span, a BIM manager (also sometimes defined as a virtual design-to-construction project manager, VDCPM for short) might be appointed. The BIM manager is retained by a design build team on the client's behalf from the pre-design phase onwards to develop and to track the object-oriented BIM against predicted and measured performance objectives, supporting *multi-disciplinary* building information models that drive analysis, schedules, take-off and logistics. Companies are also now considering developing BIMs in various levels of detail, since depending on the application of BIM, more or less detail is needed, and there is varying modeling effort associated with generating building information models at different levels of detail.

(2) *BIM in construction management*

Participants in the building process are constantly challenged to deliver successful projects despite tight budgets, limited manpower, accelerated schedules, and limited or conflicting information. The significant disciplines such as architectural, structural and MEP designs should be well *coordinated*, as two things can't take place at the same place and time. Building Information Modeling aids in *collision* detection at the *initial* stage, identifying the exact location of *discrepancies*.

The BIM concept *envisages virtual* construction of a facility prior to its actual physical construction, in order to reduce uncertainty, improve safety, work out problems, and *simulate* and analyze potential impacts. Sub-contractors from every trade can input critical information into the model before beginning construction, with opportunities to *pre-fabricate* or *pre-assemble* some systems off-site. Waste can be *minimized* on-site and products delivered on a just-in-time basis rather than being stock-piled on-site.

Quantities and *shared properties of materials* can be *extracted* easily. Scopes of work can be *isolated* and defined. Systems, assemblies and sequences can be shown in a *relative scale* with the entire facility or group of facilities. BIM also prevents errors by enabling conflict or "clash detection" whereby the computer model visually highlights to the team where parts of the building (e.g. *structural frame* and building services pipes or ducts) may wrongly *intersect*.

(3) *BIM in facility operation*

BIM can bridge the information loss associated with handling a project from design team, to construction team and to building owner/operator, by allowing each group to add to and reference back to all information they acquire during their period of contribution to the BIM model. This can yield benefits to the facility owner or operator.

For example, a building owner may find evidence of a leak in his building. Rather than exploring the physical building, he may turn to the model and see that a water valve is located in the suspect location. He could also have in the model the specific valve size, manufacturer, *part number*, and any other information ever researched in the past, pending adequate computing power. Such problems were initially addressed by Leite and Akinci when developing a *vulnerability representation* of facility contents and threats for supporting the identification of vulnerabilities in building emergencies.

Dynamic information about the building, such as sensor measurements and control signals from the building systems, can also be incorporated within BIM software to support analysis of building operation and *maintenance*.

There have been attempts at creating information models for older, *pre-existing* facilities. Approaches include referencing key metrics such as the Facility Condition Index (FCI), or using 3D laser-scanning surveys and photogrammetry techniques (both separately or in combination) to capture *accurate* measurements of the asset that can be used as the basis for a model. Trying to model a building constructed in, say 1927, requires numerous assumptions about design standards, building codes, construction methods, materials, etc., and is therefore more complex than building a model during design.

One of the challenges to the proper maintenance and management of existing facilities is understanding how BIM can be *utilized* to support a holistic understanding and implementation of building management practices and "cost of ownership" principles that support the full life cycle of a building. An American National Standard entitled APPA 1000-Total Cost of Ownership for Facilities Asset Management incorporates BIM to factor in a variety of critical requirements and costs over the life-cycle of the building, including but not limited to: *replacement of energy*, utility, and safety systems; continual maintenance of the building exterior and interior and replacement of materials; updates to design and functionality; and *recapitalization* costs.

(4) *BIM in land administration and cadastre*

BIM can potentially offer some benefit for managing stratified cadastral spaces in urban built environments. The first benefit will be enhancing visual communication of interweaved, stacked and complex cadastral spaces for non-specialists. The rich amount of *spatial* and semantic information about physical structures inside models can aid *comprehension* of cadastral boundaries, providing an *unambiguous delineation* of ownership, rights, responsibilities and restrictions. Additionally, using BIM to manage cadastral information can advance current land administration systems from a 2D-based and *analogue data environment* into a 3D digital, intelligent, interactive and dynamic one. BIM can also unlock value in the cadastral information by forming a bridge between that information and the interactive life cycle and management of buildings.

3. Anticipated Future Potential

As a new technology, BIM has the following advantages:

1) Improved visualization

2) Improved productivity due to easy *retrieval* of information

3) Increased coordination of construction documents

4) *Embedding* and linking of vital information such as vendors for specific materials, location of details and quantities required for estimation and tendering

5) Increased speed of delivery

6) Reduced costs

BIM also contains most of the data needed for building performance analysis. The building properties in BIM can be used to automatically create the input file for building performance simulation and save a significant amount of time and effort. Moreover, automation of this process reduce errors and mismatches in the building performance simulation process.

Green Building XML (gbXML) is an emerging *schema*, a subset of the Building Information Modeling efforts, focused on green building design and operation. gbXML is used as input in several energy simulation engines. With the development of modern computer technology, a large number of building performance simulation tools are available. When choosing which simulation tool to use, the user must consider the tool's accuracy and reliability, considering the building information they have at hand, which will serve as input for the tool. Yezioro, Dong and Leite developed an artificial intelligence approach towards assessing building performance simulation results and found that more detailed simulation tools have the best simulation performance in terms of heating and cooling electricity consumption within 3% of mean absolute error.

Words and Expressions

streamlined	流线型的，效率更高的	pre-existing	预先存在的
coordinate	整合，协调	accurate	精确的
visualization	可视化	utility	效用，功用
collaboration	合作，协作	recapitalization	资本重组
component	成分，组件	spatial	空间的，空间上的
dictate	支配，决定	comprehension	理解
modification	修改，修正	unambiguous	清楚的，明确的
constructability	可施工性	delineation	描绘，勾画
sustainability	可持续性	retrieval	检索
span	跨度，贯穿	embed	嵌入，植入
multi-disciplinary	多学科	schema	（计划或理论的）纲要,图解,模型
coordinated	协调的	facilities management	设施管理
collision	碰撞，冲突	across disciplines	跨学科
initial	最初的，最开始的	cost management	成本管理
discrepancy	矛盾，差异	construction management	施工管理
envisage	设想，想象	project management	项目管理
virtual	实质上的，事实上的	facility operation	设施运作
simulate	模拟，模仿	shared properties of material	材料性质共享
pre-fabricate	预制的，用预制构件组装的	relative scale	相对比例
pre-assemble	预装配	structural frame	结构框架
minimize	最小化	part number	零件编号
extract	提取，提炼，获得	vulnerability representation	脆弱性表现
isolate	隔离	replacement of energy	能源替代
intersect	相交，交叉	analogue data environment	模拟数据环境
maintenance	维修，维护		

Activities—Discussion, Speaking & Writing

Presentation

Group: 5 to 7 members

10 minutes per group (Each member should cover your part at least one or two minutes).

Clearly deliver your points of the following questions to audiences.

NEED practice (individually and together)!!

Gesture and eye contact.

Smile is always KEY!! Cover your nervousness!!

Questions for discussion and presentation

Do you know how many Green Buildings in China by now? Introduce some of them!

Writing

What's your opinion on BIM by now and in future? Write a report on the future potential of BIM.

Unit 11
International Cooperation and Exchange

Teaching Guidance

Have you ever thought about studying abroad? The purposes of studying abroad are various: improving the level of knowledge and skills, learning advanced foreign cultural knowledge, feeling foreign academic atmosphere, broadening horizons, experiencing foreign life, learning foreign languages, seeking foreign jobs, and seeking immigration opportunities.

11.1 MACE of the University of Manchester

1. You Could be Part of the University of Manchester

The University of Manchester is one of Britain's most famous and forward thinking universities, with a rich heritage stretching back 180 years and an exciting agenda for the future.

And you could be part of it…

The birth of the computer, the founding principles of modern economics, the research that led to the splitting of the atom——all these and many more world altering innovations have their roots here, at the University of Manchester. Today, it is one of the top universities for biomedical research, while its international centres exploring cancer research, world poverty, environmental sustainability and social change are producing answers to global problems that truly change lives.

The mission of University of Manchester

Its mission is to become one of the top 25 universities in the world by 2025, attracting the best students, teachers, researchers, and reputation. It's a goal that it's well on the way to achieving, backed by a major multi-million pound investment programme in facilities, staff and buildings.

Targeted by thousands of graduate recruiters, and with a thriving research community, nowhere can offer you better prospects than the University of Manchester. Decide to study here and you will be welcomed into the prestigious ranks of an institution famous for cutting-edge innovation and enterprise, situated at the heart of one of the world's most exciting student cities.

Research, Discovery and Innovation

As a postgraduate student at the University of Manchester, you'll have the opportunity to make a major contribution towards research excellence in your field. Whether studying for a taught postgraduate award, or a research degree, you will be directly involved with ground breaking research, helping to push the bound of the boundaries of creativity.

Its worldwide reputation for pioneering research and proactive relationships with industry and public services make it both a centre for academic excellence and a force for positive change. Many major advances of the 20th century began in its laboratories, such as the work by Rutherford leading to the splitting of the atom and the development of the world's first programmable computer, "The Baby", in 1948.

Today, research remains at the heart of the University. The research is aimed in a wider range of academic areas than any other UK universities and virtually all of its research has been assessed as being at international or national standards of excellence. It is of confidence of continued improvement on its impressive Research Assessment Exercise rating as it increases the number of first rate professorships, builds on its strong links to industry, and continues to invest in world class facilities.

Each year, the University attracts around 250 million of research funding from external sources, bringing the total research expenditure to almost 400 million per year and enabling to develop cutting edge research facilities, staff, programmers and discoveries. It is among the top three universities for grant funding from the main UK engineering, science and bioscience research councils.

Throughout your studies, you'll be encouraged to adopt innovative approaches to research, breaking down limitations and discovering new interdisciplinary ways of working. Thinking in a cross disciplinary way is opening up exciting new areas of study and discovery in its 23 Academic Schools and its new University Research Institutes.

2. Faculty of Science and Engineering (Overview)

With over 500 academic staff and a similar number of post-doctoral research staff, the breadth and volume of its research is unsurpassed. The research its is also world-leading in terms of quality, as shown by the results of the UK Research Excellence Framework (REF) in 2014, and by the award of the 2010 Nobel Prize for physics to two of professors in University of Manchester, Andre Geim and Konstantin Novoselov.

There are more than 70 specialist research centres and groups in the Faculty, undertaking research in fields such as photon science, electrical energy distribution systems, neural-inspired computing architectures, the structure of the universe, nuclear science and technology, tissue engineering, alternative energy sources, global and local environment studies, lightweight materials for 21st Century fuel-efficient vehicles, medical imaging, aeronautical engineering, and nano-materials.

Within the University of Manchester, we enjoy strong collaboration with colleagues in the other three faculties: Humanities, Life Sciences, and Medical and Human Sciences. Many research activities in EPS are carried out in national and international collaborations.

3. School of Mechanical, Aerospace and Civil Engineering

As one of the largest Schools in Europe that incorporates Mechanical, Aerospace and Civil Engineering, the university offers you a challenging and diverse learning environment. A degree from The University of Manchester is an international passport to success, and many students achieve excellence on a worldwide stage. Its programmes are aimed at producing top quality graduates. Hope you will rise to that challenge.

Its expertise

Its work is applicable to diverse industry sectors: aerospace, manufacturing, civil, process industries, medical, nano-engineering, energy, environment, transport and nuclear.

Its courses are amongst the best nationwide, with Civil Engineering topping the Guardian

League Tables in 2018.

The students and graduates

(1) "I would highly recommend the School of Mechanical, Aerospace and Civil Engineering to those who want to pursue a career in engineering of project management. The School has a great reputation and is globally well recognized. I have experienced first-class support with excellent learning and development opportunities."

Mehmood Alam, final-year PhD student

Discipline: Engineering Project Management

Sponsorship: Rolls-Royce, AMEC, Goodrich, EDS and The University of Manchester

(2) "After recently completing my PhD studies, I have joined ALSTOM Aerospace. I have been working at the AMF (Aerospace Manufacturing Facility), responsible for a range of design activities, including stress and aerodynamic analysis.

"My postgraduate study was funded by Airbus UK. I found the support from the company subject matter experts invaluable. During my three years of PhD studies, I learnt a lot about techniques and approaches to problem-solving that can be adapted to any engineering situation."

Kailash Sunneechurra, Senior Design Engineer ALSTOM Aerospace

(3) "After graduating from The University of Manchester EngD scheme in 2006, I joined my sponsoring organisation in a research management role, responsible for managing development project for the Ford and Honda Motor companies. I found my technical research background and the formalised management training invaluable when management large research project."

"I have recently rejoined the school of Mechanical, Aerospace and Civil Engineering to manage an executive education program for senior BP Engineering managers."

Dr Moray Kidd, Deputy Director

BP engineering Management program, School of Mechanical, Aerospace and Civil Engineering.

Entry requirements

(1) **Academic entry qualification overview**: The minimum academic entry requirements for a postgraduate program will be lower second UK Honours degree, or international equivalent in a relevant science or engineering discipline. For some programs, including our PhD and EngD research degree, the minimum academic entry requirements will normally be an Upper Second UK Honours degree, or international equivalent. Please visit the school website for further information.

(2) **English language**: You will need to be able to demonstrate competency in English language and students who do not already possess a recognized English language qualification will need to take a test such as IELTS or TOEFL and attain a minimum of IELTS 6.5, TOEFL 570 with 5.0 in the TWE. Other English language tests can be considered, including those offered by the university's language teaching centre.

(3) **Other international entry requirements**: It can accept a range of qualification from different countries. For these and general requirements, including English language, please in-

quiry entry requirements from your country.

Scholarship/Sponsorship

There are a number of funding opportunities available to applicants planning to start a taught or research postgraduate degree course at the school of Mechanical, Aerospace and Civil Engineering. Please contact to find out more.

4. Civil Engineering

Civil engineers make modern society possible. They conceive, design and build everything from roads to buildings, to bridges, to ports, to entire cities. They produce flood and sea defences; they provide water supplies; they build our railways and airports. And they do all this in a safe, sustainable and cost-effective way. Civil engineers are problem solvers who use creativity, technology, teamwork, science, and originality to provide the environment in which we live. Us your ingenuity and imagination to solve some of the world's biggest challenges appeals and read on to find out more about civil engineering education at Manchester.

The university uses academic research to directly inform students, ensuring that courses benefit from both a traditional civil engineering core and specialised expertise. Students acquire an understanding of physical sciences and principles of engineering in the natural and built environment, and are encouraged to engage with the wider society through sustainability studies and personal development programmes.

11.2 The 15th International Symposium on Structural Engineering

1. Introduction

The 15th International Symposium on Structural Engineering (ISSE-15) will be held in Hangzhou on October 24 ~ October 27, 2018. The ISSE-15 is sponsored by the National Natural Science Foundation of China (NSFC). It will be organized by Zhejiang University. The Symposium was derived from the International Symposium on Structural Engineering for Young Experts (ISSEYE) which was held in Leshan, China since 1990. It has been held biannually in China since in Harbin (1992), Shanghai (1994), Beijing (1996), Shenyang (1998), Kunming (2000), Tianjin (2002), Xi'an (2004), Fuzhou (2006), and Changsha (2008). In 2010, the title of the symposium was updated to "The International Symposium on Structural Engineering (ISSE)" and then held in Guangzhou (2010), Wuhan (2012), Hefei (2014) and Beijing (2016) with this title. Till now, it has been successfully held for 14 times. With more than 20 years' development, ISSE has become attractive to many young and middle-aged elite Chinese scholars all over the world with its distinct characteristics and features.

The ISSE-15 takes inheritance, development, openness and imagination as the tenet. The objective of ISSE-15 is again to provide a forum for experts from the research and engineering communities, who work worldwide in the broad areas of structural engineering and construction, to present recent progress in research and development; to exchange information on the topics of design theory about disaster impact, modern structural experiment, structural system and its performance, structural failure mechanism and behavior control, measurement and numerical simulation techniques, construction and maintenance management; to discuss the application of innovative technologies and techniques for the safety and sustainable development of infrastructures; to promote close international collaboration and cooperation; to figure out the future development of structures in the coming 20 years.

第十五届结构工程国际研讨会（ISSE-15）将于2018年10月24~27日在杭州举行，会议由国家自然科学基金委员会（NSFC）主办，浙江大学承办。结构工程国际研讨会（ISSE）源自1990年在乐山召开的结构工程青年学者国际研讨会（International Symposium on Structural Engineering for Young Experts- ISSEYE），该研讨会随后分别在哈尔滨（1992）、上海（1994）、北京（1996）、沈阳（1998）、昆明（2000）、天津（2002）、西安（2004）、福州（2006）、长沙（2008）召开，于2010年更为现名后分别在广州（2010）、武汉（2012）、合肥（2014）和北京（2016）召开，已成功举办14届。经过20余年的发展，该会议已经成为结构工程领域以世界华人学术骨干为主体、特色鲜明的国际学术会议。

本次举办的ISSE-15以"传承、发展、开放、畅想"为宗旨，展示学者、研究人员和广大结构工程技术人员最新研究成果；交流灾难作用、现代结构实验、结构体系与性能设计理论、结构失效机理与性态控制、实测与数值模拟技术等关键问题；研讨新工艺和新技术在社会基础设施的安全和可持续发展中的应用；推动国际间合作与交流；并畅想结构未来发展20年。

2. Themes (Major topics included but not limited to)

Structural system and design theories

Green building and industrialization

Theories and methods of structural performance analysis

Advanced structural experimental techniques

Advanced sensing technology, sensor networks, structural monitoring and control

Disaster prevention and mitigation, protection of infrastructures

Soil-structure interaction

Large scale and complex infrastructure and spatial structures

Reliability and durability of structures

Structural rehabilitation, retrofitting and strengthening

Heritage building performance and assessment

Application of artificial intelligence in Civil Engineering

会议主题（会议议题主要包括但不局限于以下范围）

结构体系与设计理论

绿色建筑与工业化

结构性能分析的理论与方法

先进结构实验技术

先进的传感技术，传感器网络，结构监控

预防和减轻灾害，保护基础设施

土-结构相互作用
大型复杂基础设施和空间结构
结构的可靠性和耐久性
结构修复、改造和加固
文物建筑的表现及评估
人工智能在土木工程中的应用

3. Participants Requirement

Participants are required to have a Doctorate degree or a senior technical title. Experts undertaking important research or engineering projects are especially welcome. Chinese and English are the two official languages used in reporting and communication during the symposium. All papers submitted to the conference are to be written in English.

参会人员需具有博士学位或具有高级技术职称的专家和学者。尤其欢迎正在主持重要研究课题或重点工程项目的专家参加。会议期间交流和报告语言可采用英语或汉语，但论文须用英语撰写。

4. Abstract

The authors should submit a two-page expanded abstract and a completed Preliminary Registration Form to ISSE-15 Secretariat through conference website (http://www.isse15.com) before the required deadline. The abstract, in WORD format, should include the paper title, abstract, key words and the information of each author such as name, affiliation, title, address, phone and fax number, and the email address. And the corresponding author should also be indicated. The abstract will be reviewed by experts in same research area. A notification of provisional acceptance and the manuscript template will be sent to the corresponding author once the abstract is accepted. All the abstracts will be published before the conference and distributed to participants during the symposium.

作者须在摘要提交截止日前，通过会议网站提交两页扩展版英文摘要（网站地址：http://www.isse15.com）。论文摘要应包括题目、摘要正文、关键词、作者姓名、作者照片、所在单位、职称、联系地址、电话和传真号码以及电子邮件地址，且须注明通讯作者。摘要将由同行专家进行评审。论文摘要通过评审的通讯作者将会收到临时录用通知和论文格式模板。论文摘要将在参会前由出版社出版、会议期间分发。

5. Proceedings

English will be the official language for papers to be included in the conference proceeding. All papers will be peer-reviewed by 2 reviewers selected from an international board of experts. The Academic Committee will gather the comments provided by peer-reviewers, and inform the corresponding author. An electronic version of the final revised paper in both WORD and PDF format should be sent to the symposium secretariat through conference website (http://www.isse15.com) before the required deadline. Minor format modification and correction may be subsequently made by the editors of the proceedings.

所有提交论文将由来自国内外的两位专家进行评审。专家评审意见由学术委员会收集，并通知相应论文通讯作者。通过评审并按评审意见修改的论文终稿电子版（包含

word 格和 pdf 格式）必须在截止日期前通过会议专用网站（http：//www. isse15. com）。论文集编辑人员保留对论文格式等做微小调整和修改的权力。

6. Hangzhou

Hangzhou's history dates back to 2200 years ago during the Qin Dynasty (221~206BC). The city was chosen as capital of the Wuyue Kingdom (907~978AD) and the Southern Song Dynasty (1127~1279AD). It is one of the seven ancient capitals of China. Hangzhou is famous for its beautiful scenery, and crowned as the paradise on earth. Hangzhou benefited from the convenience of the Beijing—Hangzhou canal and trade ports, as well as its own developed silk and grain industry, which historically served as an important commercial distribution center. Thereafter, relying on the railway, as well as Shanghai's import and export trade, its industry experienced a rapid progress. Hangzhou has a large number of cultural relics most of which scattered around West Lake, and the main representative among them include the West Lake culture, Liangzhu culture, silk culture, tea culture, etc. And a lot of stories and legendries are handed down, and become cultural representatives of Hangzhou.

杭州自秦朝设县治以来已有2200多年的历史，曾是吴越国和南宋的都城，是中国七大古都之一。因风景秀丽，素有"人间天堂"的美誉。杭州得益于京杭运河和通商口岸的便利，以及自身发达的丝绸和粮食产业，历史上曾是重要的商业集散中心。后来依托沪杭铁路等铁路线路的通车以及上海在进出口贸易方面的带动，轻工业发展迅速。杭州人文古迹众多，西湖及其周边有大量的自然及人文景观遗迹。其中主要代表性的独特文化有西湖文化、良渚文化、丝绸文化、茶文化，以及流传下来的许多故事传说成为杭州文化代表。

11.3 Events on The Hong Kong Polytechnic University

1. Shenzhen/Hong Kong Innovation Circle Research Fund for CSE Scholars

Two scholars at The Hong Kong Polytechnic University have recently been awarded a Shenzhen/Hong Kong Innovation Circle Research Fund by the Shenzhen Bureau of Science, Technology and Information. This was for a project on "Smart Wireless Sensor Networks for Living Environment Monitoring in Metropolis Area". The research project heated by DR. Y. Q. Ni of the Department of Civil and Structural Engineering (CSE) as Principal Investigator, and DR. D. Wang of the Development of Computing as Co-investigator.

This research will be the first to use deployable wireless sensor network system on interactive living environment monitoring in densely populated residential area. It aims to achieve living environment monitoring in an unsupervised, real-time, and reliable manner by making use of the smart wireless sensor network. This will be done through the integration of smart sensor nodes, wireless data transmission and communication network, as well as built-in data analysis algorithms. Researchers will first develop smart sensor nodes with capabilities of sensing, computation, storage and communication for living environment monitoring within a small region, to keep check of variables such as temperature, heat, gas, pollutant, vibration. An efficient wireless data communication and networking system will be established for the interactive communication between the smart sensor nodes which have been spatially distributed in a wide region, with the data analysis algorithms being built into each smart sensor node, analysis of monitoring data within the wireless sensor network is executed in a distributed and collaborative manner. An alarm is automatically activated by the wireless sensor network when it detects any abnormality in the living environment. The outstanding features of distributed computation capability, small volume, low ower consumption and wide area coverage make the wireless sensor network especially fit for living environment monitoring in densely populated area.

2. CSE Staff Studying Long-Range Transport of Air Pollution in Southern China

In the spring of 2009, researchers from the Department of Civil and Structural Engineer (CSCE), The Polytechnic University of Hong Kong (PolyU), carried out a large-scale field study at Mount Hengshan in Hunan Province, China. The purpose of the study was to understand the interactions between air pollution and cloud formation in the region, and the long-range transportation of air pollution between the northern and southern regions of the Chinese mainland.

A comprehensive suite of advanced instruments were deployed to the mountain site, collecting data on gases, aerosols, clouds, and rainfall.

3. Collaboration Agreement with the Institute of Urban Environment, Chinese Academy of Sciences

As the world's urban population continues to grow, it is becoming increasingly imperative to understand the dynamic interactions between human activities and the urban environment. The last few decades have seen unprecedented industrial development and urbanization in Hong Kong and the mainland in China. However, the process has also caused significant environmental problems in many regions. Both the Central Government of China and the Hong Kong Special Administrative Region Government have recognised the importance of environmental protection and sustainable development, and have given high priority to the reduction of pollution and to sustainable development.

The Chinese Academy of Science has recently collaborated with the Faculty of Construction and Land Use, The Polytechnic University of Hong Kong (PolyU) to work on the ways to facilitate technological and academic exchanges on environmental issues during rapid urbanization processes. They will collaborate in submitting research applications and to build up a joint-research laboratory on urban environmental health.

Major research areas included: Urban biogeochemistry and environmental health, Advanced water treatment technology and urban water cycle, Atmospheric chemistry and urban air quality, Utilization of solid wastes and cycle economics, Cyber city, Ecological architecture, Infrastructure and transport panning, Urban safety.

The collaboration will cover the development of academic exchange activities in various formats, and to encourage PolyU academic staff members to become visiting professors at the Chinese Academy of Sciences. This research was funded by the PolyU Niche Area Development Scheme and China's 973 Programme. Research Academy of Environmental Science, Chinese Academy of

Meteorological Sciences, and the Hunan Provincial Meteorological Bureau participated in the experiment. Professor Tao Wang from CSE, PolyU, led the study.

4. A General Donation From Prof. K. K. Choy to CSE

The Department of Civil and Structural Engineering (CSE) of the Hong Kong Polytechnic University (PolyU) has for many years benefited from the support from Prof. K. K. Choy, Assistant Director of the buildings Department. Recently, Prof. Choy has established the K. K. Choy Scholarship with a donation of HKD800000 for CSE. The Scholarship with a donation aims to encourage CSE students to excel in their academic pursuits, and to nature talents in the field of civil and structural engineering.

Prof. Choy is an alumnus of CSE, graduating with an Associate ship in Civil and Structural Engineering in 1975. He has been appointed as an Adjunct Professor by PolyU and is a Council Member of the Institution of Structural Discipline Advisory Panel, the Hong Kong Institution of Engineerings. He is also chairman of the Board of Education and Research, the Hong Kong Professional Green Building Council.

5. CSE Professors Working for Blue Skies Across Asia

On 19 April 2011, Prof. Shuncheng LEE of the Department of Civil and Structural Engineering (CSE) gave a presentation on Hong Kong in air quality management in Jakarta at the Trsakti University. The audience were about 100 people, including academics and Indonesian government officials, whose overwhelmingly positive responses have provided the impetus for the creation of an Environmental Training Centre by CSE to train young people in partner countries.

This visit to Jakarta is part of the Blue Sky Asia - Clean Air Exchange Programme that is funded by Fredskorpset, also known as FK Norway, a Norwegian governmental body under the auspices of the Norwegian Ministry of Foreign Affairs. Its leading partner is the Clean Air Initiative for Asian Cities Center (CAI-Asia). Besides PolyU's CSE, its other partners are the Clean Air Network Nepal, Clean Air Sri Lanka, Komite Penghapusan Bensin Bertimbel of Indonesia, as well as the Vietnam Clean Air Partnership of the Vietnam Association for the Conservation of Nature and Environment.

Since the exchange programme started in September 2008, PolyU has been able to conduct research on the air quality of many Asian cities, including Hanoi, Kathmandu, Columbo and Jakarta. Under the leadership of Dr Wing Tat HUNG, the programme has enabled the exchange of postgraduates amongst the partner countries to work for 10 months monitoring and managing air quality. CSE has provided all the technical support and relevant training. Other CSE academics involved in this programme are Dr GUO Hai and Prof. LEE, who also attended the programme's annual review meeting on his recent trip to Jakarta. Under this exchange programme, a memorandum of understanding has been signed with the Department of Mechanical Engineering of the University each year, a comparative air quality study has been conducted in the city where the participant is from. Since the current participant, Agung Muhammad, from Indonesia, a comparative study is underway in Jakarta now.

Unit 12
Preparation for Being a Civil Engineer

Teaching Guidance

Opportunities Favor the Prepared Mind.
——Louis Pasteur

Today is made for us. Every minute counts. Today's decisions are tomorrow's realities. Be optimistic! Be determined! Those who hesitate are lost. Every man is a king!

The first step to be a civil engineer is generally to study civil engineering in a university or college, or major in civil engineering or other related programs. In most countries, the certificate of Registered Engineer is only given to those who have accepted higher education in accredited programs. In this chapter, the reader will acquire the information about the typical content of these programs.

12.1 What Kind of Knowledge is Necessary for a Civil Engineer?

Engineering education in universities domestic and abroad includes general education and special engineering education. At first, science and mathematics should be mentioned in general education. Engineering is a system of the applying of science and technology, so scientific principles set the foundation of engineering. This is the most important difference between modern civil engineering and ancient construction activities, although construction has always depended to some extent on scientific principles. Since the Industrial Revolution, and even as far back as the Renaissance, civil engineering has always been a branch of technologic science. For these reasons, science and mathematics become the common base of engineering education including civil engineering education.

Owing to the accumulation of several centuries, modern science has accumulated a massive body of literature and knowledge. However, the beginner needs not sit under an apple tree to discover the laws of universal gravitation as Isaac Newton did in legend. Neither does he have to exhaust his brain for the principle of transform between energy and the mass. Based on the work of the numerous pioneers, new students can now enter into the paradise of science easily. Nowadays, engineering is a synthetic system not only depending on traditional mechanics, but also closely relating to advanced science. You can find the courses such as physics, chemistry, computer science, material science, environmental science, and perhaps more, in your civil engineering program.

One characteristic of modern science is that it can be described exactly and beautifully by mathematics. So the engineer should grasp this powerful tool to solve the problems they will meet in engineering analysis, design, planning and control. In this aspect, engineering students should learn advanced mathematics including analytic, geometry, differential and integral calculus, probability, numerical equation. In addition, study of linear algebra, matrix, probability, numerical methods is usually required by Civil Engineering Program. Using all of this knowledge, an engineer can tell whether a house or a bridge is safe or dangerous when earthquake occurs, or when it is hit by a hurricane. How can the skilled engineer do it? The engineer does this by using abstract models from physical objects, which can be described and predicted by mathematics. Mathematics provides engineers with a solid foundation in their engineering activities. Furthermore, by strict training through verification, deduction and calculation in the study of mathematics, one will accustom oneself to logicality, strictness, and more rationality, which are important qualities for a good engineer.

An engineer not only just takes the responsibility or the technology and production activities of a project but also has the duty to the society. Does your engineering project benefit your people and society or harm them? A qualified engineer should be conscientiously aware of this point at all times and for this reason, universities also organize social science and humanities education for their students. Students enrolling in engineering programs should accept the education in this aspect. Philosophy, ethics, history, literature, aesthetics, as well economics, management and foreign language are a useful and necessary tool.

The necessary knowledge for professional occupation of civil engineering is composed of two parts: base knowledge for entire civil engineering and corresponding knowledge for a special aspect.

Most civil engineering projects can be seen as varieties of structures. In order to ensure the safety of structures, civil engineers should understand their mechanical properties, such as forces, stresses, displacements and deformations of the structures, caused by the weight of structure itself and facilities, winds on the structure, vehicles, varying of temperatures, and perhaps earthquakes. Courses, usually named mechanics of materials, structural analysis, elasticity, are set for this purpose. Because civil engineering projects are laid on or under the ground, to know soil and rock properties well is necessary. Thus geo-engineering, soil mechanics and foundation studies are also base knowledge. Water and wind, those will act on or react with the structures, have common properties in the view of mechanics, and fluid mechanics deals with the concerned theories. Furthermore, a knowledge of engineering chart drawing (a skill to express the design idea by pictures in common rules understood by engineers, technicians and workers), surveying (to measure the landform for construction), and electricity, machinery, construction management and general technic, budget, bidding the tendering are also required.

Since civil engineering covers many fields of knowledge with many aspects, it is impossible to learn all of the knowledge in these areas. Almost all of the universities in the world provide students with several options to enable them to specialize in the fields. Such a method is also being re-accepted by civil engineering education in China since 1998, although it was the way in the early history of higher civil engineering education before 1950s. For example, students can now choose options in building structures, bridge, tunnel, road pavement and construction, railway and so on, to know how to design, construct, and organize a civil engineering project. And the students are usually encouraged to choose more options for their future professional life.

12.2　What Can the University Education Provide for Students?

COURSES: Basically, university offers students a variety of courses. The branches of knowledge mentioned above are involved in the courses and courses are usually divided into three types: requirement, approved electives and free choice.

The requirement and approval electives are both the courses that the students majored in must learn. There are some differences between the two types. Students cannot miss any requirement course while have limited right to elect some of the approved electives. In that case, a university usually tells students the minimum which they should choose in the list of the approval electives. As for the free choice, universities normally ask for a necessary number of credits or class hours. Those who hope to graduate and be awarded the corresponding degree, have to meet their aquirement of the university or the school.

Universities should continually adjust teaching plans and course tables with the development of science and technology, to meet the needs of future engineers. So the contents of courses are changed from time to time.

TEACHERS: As in middle and high schools, teachers in universities give lectures, check homework, organize panel discussion for special problems, guide the students to experiments and also check answers in test sheets at the end of semester. Simultaneously, most of them play the role of scientists and/ or engineers. They publish research papers in journals, spend much time in laboratory to verify a new discovery, test a renewal material with the engineering purpose. Some of them are registered engineers if their field is civil engineering, and even have their own design institutes. In famous universities, when you knock a door to ask your professor a question, you will be probably told that the professor who you are talking is a respectable academician of Academy of Science or Engineering. The groups of wisdoms, who are good at theories and practice experience, are the best gift the universities afford to the students. Unlike the teachers in middle and high schools, university professors rarely monitor your daily study, because they appreciate students should study on individual initiative.

With the development of internet, the tele-course is becoming fashionable. A young student will be in a puzzle about the large number of teachers in one university, but will find, face to face lecture and discussion is always charming, and by means of direct communication not limited in speech. Communication is also by means of expression of teacher's eyes and gestures. The close distance between you, your classmates and the lecturer, will make for an excited atmosphere. It is why since Socrates, Confucius, no matter how modern the society has become, and no matter what kind of high tech is introduced into the education process, the university always keeps its campus and excellent scholars in a remarkable size.

LABORATORY AND SITE PRACTICE BASE: For engineering colleges, the laboratories equipped with variety of test machines and measuring devices, and opened to students are indispensable. There are several types of experiments with special purpose, for demonstration, observation, validation, practical training, exploration, or others. The basic experimental skill necessary for engineers can be learned in the laboratories. Most of test items are specified in the textbook, and detailed instruction is printed. In recent years, universities in China encourage students to design the experiments themselves, and do what they are interested in the related fields, to make students have the desire for innovation.

It is cognized that a qualified engineer should possess rich experience obtained from engineering practice, so practical exercise becomes one important part of the education plan of civil engineering program. Laboratory training is part of this practical training. Others are design work both in classroom and in workroom of consulting companies or design institutes, construction site work, geologic investigation, surveying and measurement outside. In most cases, Chinese universities set practice bases at construction companies and design institutes. Usually students are requested to join the construction site work during the summer or winter vacation. A new procedure is tried in a few universities to ask students to search the projects being constructed and go there for their practical training. The procedure itself is taken as a practice. Most universities take the practical trainings to be requirement or approved electives.

LIBRARY AND OTHER INFORMATION SOURCES: Self-study is a typical mode of university students, successful students are always those who do not satisfy the contents of lectures and homework given by teachers. For them, comprehensive reading is undertaken outside indicated textbooks. Books, journals, reports and dissertations in the form of collection of printed pages which are stored in the book shelves are also read. Of course, the libraries in modern universities are reformed with the computer system and network, and the electronic libraries make it more convenient for students more convenience to borrow and read. The ability to search, find and grasp information becomes more and more important in this age, and it is the task of the university education to let students have this ability.

SPIRIT AND ATMOSPHERE: In the common sense, the universities are the place where there are freedom for thinking, equality in academy, and advocation of creation. Furthermore, the alternation of new students every year, make university campuses full of the energy of the younger generation.

ACTIVITIES OUT OF CLASS: There are different student organizations in the campus that help connect classroom to career, develop professionalism, increase technical proficiency and refine ethical judgment. For example, the Institute of Civil Engineering (ICE) of British welcomes the students enrolled in program of civil engineering to be student members; even ICE develops its members in Chinese universities. Recently American Society of Civil Engineering (ASCE) joined this action too. There are many sports teams for soccer, basketball, badminton, swimming, track and field, which are organized inter-class, department and even college. Societies in literatures and arts, will afford students a total different area from those in the class.

12.3 What Abilities Shall a Future Civil Engineer Possess?

THE ABILITY TO APPLY THE KNOWLEDGE: Elementary knowledge is essential to a civil engineering student. In common, by four-year period study, the student should be proficient in mathematics through differential equations, probability and statistics, calculus-based physics, and general chemistry; proficiency in the material mechanics, fluid mechanics, structural analysis and geo-tech knowledge; good command of the primary skills for engineering survey, drawing, test, and calculation and design, and at least deeper understanding in several major civil engineering areas.

The emphasis should be shifted to the application of the knowledge after we understand the importance of the knowledge. 'To know', is mere the first step. For engineers, the more important thing is to apply his knowledge, i.e. natural science, mathematics and elementary engineering knowledge recorded in the textbooks or papers in the form of rules, principles, formulae and data, to solve engineering problems.

THE ABILITY TO CONDUCT EXPERIMENTS AND EXPLAIN THE RESULTS: The ability to plan and conduct experiments and analyze the results are basic aspects of the engineers' abilities.

The future engineer is required to conduct laboratory experiments and to critically analyze and interpret data. Though many problems can be solved efficiently and economically by computation in a fine mechanical model, it is not everything. When new material or new structural system is used in civil engineering project, there are new variables which are not reflected, covered in the ready-made model. It will be dangerous if engineers do not change their mathematical model in time. However, how to calibrate the model? The most practical way is to do an experiment. Similar things also encounter in built-up or 'older' constructions, because there are many unknown factors. For example, material used in the structure will weaken, be damaged and lose its function through the duration of a structure's life while the change cannot usually be fully expected at the beginning. And on the other hand, the surroundings, conditions and real loads can also change. Engineers and researchers make the same phenomena, in most cases, to recur in the laboratory, so that they can reveal the mechanism which now should be understood for the purpose of the safety of the structure. According to the base theory, research engineers are able to judge the results of the experiments, it is common that the observed phenomena or obtained data in the experiments conflict with the known knowledge. In this case, the conflict will bring new discovery and improve an engineer's work. Giving a rational explanation to a seemingly strange phenomenon is a wonderful task. It needs to synthesize knowledge of many subjects and to create new knowledge which is not mentioned or recorded in the literature.

THE ABILITY OF DESIGN: For engineers, the ability to design a system, a component, or a procedure of construction is basically required. Civil engineers are creating substantial entities every day and everywhere in the world. Before they make them, they should be described. It is the description of the non-existed entity that is called 'design'. The design shows what the future project is, and how to make it in a language which can be understood by constructors. The engineering design is quite different from the design of a piece of artwork, though we sometimes hear the admiration for a building as 'a graceful sculpture'. However, an artist can make a sculpture horse supporting only by one hoof, it will be impossible at ten times the size because the weight increases in three power of size. Here the key factors will be functionality, safety and low-cost. It means that only the design which meets these requirements is practicable. So the engineering design work should obey the design codes, specifications and guidance which are based on scientific principles and the summary of accumulated experience. On the other hand, as an enterprising engineer, he or she never satisfies the existed ways or technics, so to search a possible way under the limited conditions to realize the 'impossible' things in design will be a challengeable but charming work forever.

THE ABILITY TO COOPERATE WITH WORKING TEAM: An engineer never works alone. Each project is a system, so the design work involves many people's efforts. For a big size building structure, the structural engineers should work with other experts from different disciplines, such as architects, surveyors, mechanical engineers and electricians. In the past, a skilful engineer would play several roles in a project with small size, but nowadays the different jobs should be taken by qualified engineers possessing certificates. When you are in the position of chief engineer in the work team, you should be in more harmony with your fellows. In order to cooperate with others well, every engineer should know how to hear and understand others, to consider things in both sides, you and your fellows, and to make necessary concession after discussion or even quarrels.

THE ABILITY TO COMMUNICATE EFFECTIVELY: This is the ability which is an engineer should pay more attention in the modern society. To the engineer, as a designer, you should let your clients to accept your design, recognize that what you design is the most suitable one in many possibilities; you should let the examiners and officers from government believe that your design accords with the law and specification so that the public and surroundings are safe; and you should let the contractors, manufacturers and construction companies understand your consideration and its rationality and feasibility. After being an engineer, you will find that you are frequently asked to attend meeting, to explain something for the project you designed, and you have to go to the construction site to hear new problems and tell the technician the answers. All of these need good communication with others.

Unfortunately, until now our high schools and universities gave few chance to most of students to train their communication skills. Young engineering students now should take this seriously, and make communication ability.

The basic element of communication is to speak. So try to look on your audience no matter in seminars, in meetings or even in your friends' parties, make your voice loud, speak clearly and use plain but vivid vocabulary as possible.

Besides speaking, the effective communication includes writing skill and expression of one's idea both in pictures and simple formulas. There are many skills you should learn, but the most important thing is to remember that the purpose of effective communication is for thoroughly understanding between you and your companion.

Communication is not the same thing performing on a stage, where the key point should be exchange of information successfully. Good communication skill also includes hearing and considering companion's opinion. Discussion is also involved in the process of communication.

12.4 How to Match the Demands of the Program Education?

Through four-year-period study to make yourselves acquire basic knowledge and training for being an engineer is one of the main targets that makes you enter into a university and enroll into a special engineering program.

University is a new circumstance to freshmen students. For those who just left university and perhaps many of them are first time to sleep in dormitory of school, they should be familiar with the new life as early as possible.

STUDY IN CLASSROOM: Needless to say, study is the most important task. There are many things to 'study'. However, to study and understand the knowledge which are necessary for the education objectives as introduced previously and specified by the program, education plan is the basic requirement.

As a student, you have had the school experience more than ten years, so you know the study skill well: reading textbooks, attending the lectures, taking notes when listening, doing homework…, those are almost the same as in high schools. But something has changed.

The engineering students usually do not have their fixed classroom. They should move from one building to another during the ten-minute break between classes. Nobody shares one standard curriculum schedule with his classmates in the same program, especially in the junior and senior year. Students have the opportunity to choose what they 'prefer', and every one shall type the number of the courses he wants to join in the next semester into the computer registered system, or after a long queue outside the administration office to fill in course register form. To Chinese students, the most difference from the traditional high school is perhaps that no teacher will strictly monitor your daily study life.

Are you free? Certainly. But, just to certain extent.

Same as the other programs, the Civil Engineering Program requires necessary credits according to the class hours and the importance. After passing the examination, you can obtain the pointed credits. If the program asks its students to fulfill total 150 credits, you will never expect to be awarded the engineering bachelor degree in the case that you earned only 149 credits! Furthermore, as you have known, the courses you have to take are divided into three types, requirements, approval electives and free choice, but to each of the three types, the program education plan specifies a certain amount of credits you have to obtain. That is to say, your freedom is not infinite.

Sometimes, a student will be informed that he did not meet the requirement of the program because he does not pick enough credits in approval electives indicated by the program education plan. So, students had better to read program education plan and student manual carefully once enrolling in the program, and to follow it in the following days. What your tutor who is designated by the department for you, if any, can do is to give you some suggestions or advice when you consider to choose something.

To finish all the courses, the program task is important, and to get high point is also encouraged. When you pursue advanced degree study, or apply for a good position in your career life after graduation, high points are always helpful at the beginning. However, good students are not those who only know the description printed on the books or recite the formulae, but fail to explain practical phenomenon, to discover unknown things and to have strong motivation to create new knowledge himself. So university professors encourage students to consider problems in different views, and appreciate students to observe in their own eyes and to ask questions after thinking.

JOIN ACTIVITIES IN CAMPUS OUT OF CLASS ACTIVELY: Since an engineer needs to learn effective communication with others and smooth cooperation with work teams, and to be a good fellow and a successful leader both in engineering and social activities, engineering students ought not to localize their 'study' only in academy or pure specialty. Fortunately, a university is such a school that provides with plenty of opportunities to those who would like to develop their multi-talents so that campus activities are called the 'amateurish classroom'. To join one or several activities which attract you in variety of campus activities, i.e. sports, drama and concerts, forums, competitions, clubs and reading party, will benefit your spirits and brain, enlarge your friend circle and get a way more comfortable to develop yourself. It is the university tradition to encourage students to join campus activities.

PERSIST IN PHYSICAL EXERCISE: It is not a special requirement to civil engineering program students. Keeping in good health makes people to have confidence to live and work, to ensure the engineers energetically devote themselves into heavy work. By the way, though it is said not to work too heavy, in fact the work of civil engineers is really a heavy one, considering the duty, engineer must take for the safety of human being and the society!

Universities seek two main achievements in this aspect: to let the daily physical exercise become one of the personal customs of students and to train students to have some basic skills for physical exercise. Both of these are indispensable preparation for a qualified engineer.

BE AWARE OF SOCIAL RESPONSIBILITY: Why has society established a register engineer system, and why has this system been widely accepted by most of the industrial countries? The answer is that each engineering project that engineers involved in is not only a solution to a pure technical problem. At first, it will relate to the safety of life and estate. The failure of a building, collapse of a bridge or even a serious accident when undergoing construction may induce a real catastrophe to people, and make the loss of life and estate. So society asks that engineers who take the technical responsibility to the projects must be those who are qualified in knowledge and abilities. The procedure to recognize the candidate's qualification in engineering is the matter of register engineer system.

With the development of natural and social science, people have more comprehensive understanding to human being and the relation with the world. In such a background, engineers should consider more and take larger responsibilities. The engineers are being required to understand the relation of his engineering projects with the society, and the influence of the projects on environment and continuous development. For example, if an industry building to be built will bring high benefits to investors, but also high risk to pollute the rivers and surrounding soils, what should the civil engineers do? The civil engineers shall be aware of the responsibility to cooperate with the experts in that field to solve the problem. In that case, a structural engineer may adjust the previous concept design if necessary.

To be a responsible and conscientious engineer, the engineering student in the university should leave himself enough time to contact comprehensive knowledge about ethics, history and cultures of the different construction regions, beyond engineering subjects. The student needs to develop fine personality. A selfish person will be difficult to be a good engineer.

12.5 It's Never Too Late to Learn

Remember that in this chapter we only described the first step to be a qualified engineer as the author declared at the beginning. If you read the document of your university, in the section about the program education objectives, usually you can read the sentences that one of the objectives is to make students accept the primary training for being an engineer. Of course when you go through the first step successfully, that is to say, you graduate from the university and awarded the Bachelor Degree of Engineering, you will march towards to the second target to be a registered engineer. You should practice your special knowledge learned in university under the guidance of the elder registered engineers who are your colleagues or heads in your institute, company or laboratory. You do the real engineering design and construction, as assistant engineer at first in the usual case, and then accumulate experience by yourself, make familiar with the related laws, codes, rules, specifications and common manners. After four or five years, depending on your achievements, you will have the opportunity to acquire the quality of registered engineer.

In the age of 'knowledge blast', engineers have to renew their knowledge with the progress of science and technology step-by-step. So, to be an engineer, it is necessary to study continuous study all your life using the skills given to you by university.

Further Reading

The following is an example of responsibility description for project managers.

Project Management Responsibilities

POSITION:　　　　　*Project Manager*
REPORT TO:　　　　*Managing Director and Management Committee*
SUBORDINATES:　　*All Project Department Staff*

Responsibilities

Project management responsibilities are shown as follows:

1. Responsible for the management, control and smooth operation of the project division.

2. To ensure that during project implementation, all potential problem areas are identified and all issues are addressed immediately with clients.

3. To ensure that all clients' requirements and specifications in tender and quotation are met.

4. To ensure that all client requests are attended promptly and all tenders and quotations are competitively priced and promptly communicated to the clients.

5. To ensure that all undertakings or projects in hand are efficiently and effectively monitored to maximize profitability and in accordance to the contractor's requirements.

6. To ensure development of productivity, safety, quality and cost improvement strategies.

7. To ensure proper distribution of work and responsibilities daily to sub-ordinates.

8. To ensure division operation meets local environment and safety regulations/requirements.

9. Managing correspondence with consultants.

10. To liaise with the account division to ensure that cash flow is sufficient.

11. To review, recommend and approve staff promotion.

12. To ensure that all training needs of the personnel are met.

13. To review hiring of new recruits for non-executive in own division.

Authority

Approve the purchase requisition document, leave application, overtime application and any other official documents authorized by project manager.

Pre-construction Stage

1. To lead a project team members to produce a report on overall planning of the project involved. The report should be comprehensive and should cover the followings: a) Master programme. b) Progress report. c) Site organization. d) Site layout plans (location of site office, Workers' quatters, Tower Crane Location, catch-platform, Requirement for highrise, etc). e) Project methodology. f) Organizing project sample submission. g) Site survey & Site dilapidation report. h) Materials & Machineries requirement. i) Application of utilities, e.g. Water, electricity telecommunication. j) Submission of authority requirement, e.g. JKKP Form 103, CIDB, etc. k) Safety & Health plans which includes hazard identification, risk assessment and risk control (HIRARC).

2. To conduct initial meetings with immediate involvement sub-contractor to produce the following reports: Man-power planning, Materials planning, Machineries planning.

3. To liaise with contract department to produce the following reports: Cost Plan, Insurance,

Initial Materials Purchase, Initial Machineries rental.

Construction Stage

During the construction stage, the company would expect the Project Manager to perform the following tasks.

1. Project Division

a) Project briefing to the site team, which shall include company goals, project goals and project methodology. b) Managing site activities. c) Monitoring materials & machineries usage. d) Implementation of company's quality system, Standard Operating Procedures (S.O.P) & Safety Plan. e) Liaising project activities between client, consultants and site team. f) Fortnight report to management on work progress. g) Attending Project Meetings / Client Consultant Meetings (CCM) and submission minutes of such meeting for management perusal. h) Organizing site meetings and producing such minutes' meetings for management perusal. I. Site Staff Meetings; II. Project Co-ordination Meetings with Domestic & Nominated Sub-Contractor, which includes M&E Coordination Meetings and Architectural Coordination Meetings. i) To attend Higher Management Meetings and brief board of directors of project status. j) Constantly review and monitor progress of works according to Master Programme & Producing Micro Programme for works delayed.

2. Contract Division

a) Liaising with Contract Manager to produce monthly cost plan report for management perusal.

b) Monitoring Material purchase and Machineries rental with contract manager / purchasing/ store.

c) Certification of Domestic Sub-Contractor work done.

Post-construction Stage

During the post construction stage, the company would expect the Project Manager to perform the following tasks.

1. Project Division

1) Liaising with consultant and client management team in: a) Testing & Commissioning. b) Submission of Operation Manual. c) Submission of As-Built drawings.

2) Liaising with consultant & local authority on all approval matters such as Inspection, JPA Inspection, etc.

3) Appointing Supervisor to carry out duties on Defects Liability Period (DLP) and proper documentation.

4) To monitor touch-up items on DLP period to ensure defect items is addressed according to contract document.

5) To liaise with consultant on end of DLP.

2. Contract Division

a) Liaising with Contract Manager to produce debit notes on defect items. b) Liaising with

Contract Manager to implement proper hand – over of documents for client & consultant submission. c) To assist Contract Manager on Post-Mortem on each project hand-over for improvements.

Activities—Discussion & Speaking

Presentation

> Group: 5 to 7 members
> 10 minutes per group (Each member should cover your part at least one or two minutes).
> Clearly deliver your points of the following questions to audiences.
> NEED practice (individually and together)!!
> Gesture and eye contact.
> Smile is always KEY!! Cover your nervousness!!

Questions for discussion and presentation

Preparation for being a civil engineer:

1. What kinds of knowledge are necessary for a civil engineer?

2. Are you able to a general description for the necessary knowledge that provides a future civil engineer with solid science and engineering basement?

3. What can the university education provide for students?

4. What abilities shall a future civil engineer possess?

5. How do you match the demands of the program education?

6. Why does an engineering student have to have good understanding for human science and the knowledge relating non-engineering fields?

Appendix
References & Acknowledgements

Appendix 1: General Terms
附录1：土木工程一般术语

1. 土木工程 — civil engineering
2. 工程结构 — building and civil engineering structures
3. 工程结构设计 — design of building and civil engineering structures
4. 房屋建筑工程 — building engineering
5. 建筑物（构筑物） — construction work
6. 结构 — structure
7. 基础 — foundation
8. 木结构 — timber structure
9. 砌体结构 — masonry structure
10. 钢结构 — steel structure
11. 混凝土（砼）结构 — concrete structure
12. 特种工程结构 — special engineering structure
13. 房屋建筑 — building
14. 工业建筑 — industrial building
15. 民用建筑 — civil building; civil architecture
16. 构件 — member
17. 截面 — section
18. 梁 — beam; girder
19. 拱 — arch
20. 板 — slab; plate
21. 壳 — shell
22. 柱 — column
23. 墙 — wall
24. 桁架 — truss
25. 框架 — frame
26. 排架 — bent frame
27. 刚架（刚构） — rigid frame
28. 简支梁 — simply supported beam
29. 悬臂梁 — cantilever beam
30. 两端固定梁 — beam fixed at both ends
31. 连续梁 — continuous beam
32. 叠合梁 — superposed beam
33. 桩 — pile
34. 板桩 — sheet pile

35.	连接	connection
36.	节点	joint
37.	伸缩缝	expansion and contraction joint
38.	沉降缝	settlement joint
39.	防震缝	aseismic joint
40.	施工缝	construction joint

Force and Action
力和作用

1.	线分布力	force per unit length
2.	面分布力	force per unit area
3.	体分布力	force per unit volume
4.	力矩	moment of force
5.	永久作用	permanent action
6.	可变作用	variable action
7.	偶然作用	accidental action
8.	固定作用	fixed action
9.	自由（可动）作用	free action
10.	静态作用	static action
11.	动态作用	dynamic action
12.	多次重复作用	repeated action; cyclic action
13.	低周反复作用	low frequency cyclic action
14.	自重	self weight
15.	施工荷载	site load
16.	土压力	earth pressure
17.	温度作用	temperature action
18.	地震作用	earthquake action
19.	爆炸作用	explosion action
20.	风荷载	wind load
21.	风振	wind vibration
22.	雪荷载	snow load
23.	吊车荷载	crane load
24.	楼面、屋面活荷载	floor live load; roof live load
25.	作用组合值系数	coefficient for combination value of actions
26.	作用效应	effects of actions
27.	作用效应系数	coefficient of effects of actions
28.	轴向力	normal force (axial force)
29.	剪力	shear force
30.	弯矩	bending moment

31.	双弯矩	bi-moment
32.	扭矩	torque
33.	正应力	normal stress
34.	剪应力	shear stress; tangential stress
35.	主应力	principal stress
36.	预应力	prestressed
37.	位移	displacement
38.	挠度	deflection
39.	变形	deformation
40.	弹性变形	elastic deformation
41.	塑性变形	plastic deformation
42.	外加变形	imposed deformation
43.	约束变形	restrained deformation
44.	线应变	linear strain
45.	剪应变	shear strain; tangential strain
46.	主应变	principal strain
47.	作用效应组合	combination for action effects
48.	作用效应基本组合	fundamental combination for action effects
49.	作用效应偶然组合	accidental combination for action effects
50.	短期效应组合	combination for short-term action effects
51.	长期效应组合	combination for long-term action effects
52.	设计限值	limiting design value

Strength and Failure 强度和破坏

1.	抗力	resistance
2.	抗压强度	compressive strength
3.	抗拉强度	tensile strength
4.	抗剪强度	shear strength
5.	抗弯强度	flexural strength
6.	屈服强度	yield strength
7.	疲劳强度	fatigue strength
8.	极限应变	ultimate strain
9.	弹性模量	modulus of elasticity
10.	剪变模量	shear modulus
11.	变形模量	modulus of deformation
12.	泊松比	Poisson ratio
13.	承载能力	bearing capacity
14.	受压承载能力	compressive capacity

15.	受拉承载能力	tensile capacity
16.	受剪承载能力	shear capacity
17.	受弯承载能力	flexural capacity
18.	受扭承载能力	torsional capacity
19.	疲劳承载能力	fatigue capacity
20.	抗裂度	crack resistance
21.	极限变形	ultimate deformation
22.	稳定性	stability
23.	空间工作性能	spatial behavior
24.	脆性破坏	brittle failure
25.	延性破坏	ductile failure
26.	抗力分项系数	partial safety factor for resistance
27.	材料性能标准值	characteristic value of a property of a material
28.	材料性能分项系数	partial safety factor for property of material
29.	材料性能设计值	design value of a property of a material
30.	几何参数标准值	nominal value of geometric parameter

Appendix 2: Canonical Terms
附录2：土木工程规范术语

一、*Load code for the design of building structure*
《建筑结构荷载规范》 GB 50009—2012

1. 永久荷载	permanent load
2. 可变荷载	variable load
3. 偶然荷载	accidental load
4. 荷载代表值	representative values of a load
5. 设计基准期	design reference period
6. 标准值	characteristic value/nominal value
7. 组合值	combination value
8. 频遇值	frequent value
9. 准永久值	quasi-permanent value
10. 荷载设计值	design value of a load
11. 荷载效应	load effect
12. 荷载组合	load combination
13. 基本组合	fundamental combination
14. 偶然组合	accidental combination
15. 标准组合	characteristic/nominal combination
16. 频遇组合	frequent combination
17. 准永久组合	quasi-permanent combination
18. 等效均布荷载	equivalent uniform live load
19. 从属面积	tributary area
20. 动力系数	dynamic coefficient
21. 基本雪压	reference snow pressure
22. 基本风压	reference wind pressure
23. 地面粗糙度	terrain roughness
24. 温度作用	thermal action
25. 气温	shade air temperature
26. 基本气温	reference air temperature
27. 均匀温度	uniform temperature
28. 初始温度	initial temperature

二、Design Code of Concrete Structure
《混凝土结构设计规范》GB 50010—2010

1. 素混凝土结构   plain concrete structure
2. 普通钢筋   steel bar
3. 钢筋混凝土结构   reinforced concrete structure
4. 装配式混凝土结构   precast concrete structure
5. 装配整体式混凝土结构   assembled monolithic concrete structure
6. 叠合构件   composite member
7. 深受弯构件   deep flexural member
8. 深梁   deep beam
9. 先张法预应力混凝土结构   pretensioned concrete structure
10. 后张法预应力混凝土结构   post-tensioned prestressed concrete structure
11. 无粘结预应力混凝土结构   unbonded prestressed concrete structure
12. 有粘结预应力混凝土结构   bonded prestressed concrete structure
13. 结构缝   structural joint
14. 混凝土保护层   concrete cover
15. 锚固长度   anchorage length
16. 钢筋连接   splice of reinforcement
17. 配筋率   ratio of reinforcement
18. 剪跨比   ratio of shear span to effective depth
19. 横向钢筋   transverse reinforcement
20. 预应力筋   prestressing tendon and/or bar
21. 预应力混凝土结构   prestressed concrete structure
22. 现浇混凝土结构   cast-in-situ concrete structure

三、Technical specification for concrete structures of tall building
《高层建筑混凝土结构技术规程》JGJ 3—2010

1. 高层建筑   tall building; high-rise building
2. 房屋高度   building height
3. 框架结构   frame structure
4. 剪力墙结构   shearwall structure
5. 框架-剪力墙结构   frame-shearwall structure
6. 板柱剪力墙结构   slab-column shearwall structure
7. 筒体结构   tube structure
8. 框架核心筒结构   frame-corewall structure
9. 筒中筒结构   tube in tube structure
10. 混合结构   mixed structure; hybrid structure
11. 转换结构构件   structural transfer member

12.	转换层	transfer story
13.	加强层	story with outriggers and/or belt members
14.	连体结构	towers linked with connective structure (s)
15.	多塔楼结构	multi-tower structure with a common podium
16.	结构抗震性能设计	performance based seismic design of structure
17.	结构抗震性能目标	seismic performance objectives of structure
18.	结构抗震性能水准	seismic performance levels of structure
19.	异形柱	specially shaped column
20.	异形柱结构	structure with specially shaped columns
21.	柱截面肢高肢厚比	ratio of section height to section thickness of column leg

四、*Unified standard for reliability design of building structures*
《建筑结构可靠度设计统一标准》GB 50068—2018

1.	可靠性	reliability
2.	可靠度	degree of reliability (reliability)
3.	失效概率	probability of failure
4.	可靠指标	reliability index
5.	基本变量	basic variable
6.	设计基准期	design reference period
7.	设计使用年限	design working life
8.	极限状态	limit state
9.	设计状况	design situation
10.	功能函数	performance function
11.	概率分布	probability distribution
12.	统计参数	statistical parameter
13.	分位值	fractile
14.	作用	action
15.	作用代表值	representative value of an action
16.	作用标准值	characteristic value of an action
17.	组合值	combination value
18.	频遇值	frequent value
19.	准永久值	quasi-permanent value
20.	作用设计值	design value of an action
21.	材料性能标准值	characteristic value of a material property
22.	材料性能设计值	design value of a material property
23.	几何参数标准值	characteristic value of a geometrical parameter
24.	几何参数设计值	design value of a geometrical parameter
25.	作用效应	effect of an action

26. 抗力　　　　　　　　　　　　resistance

五、Code for acceptance of construction quality of concrete structures
《混凝土结构工程施工质量验收规范》GB 50204—2015

1. 混凝土结构　　　　　　　　　concrete structure
2. 现浇混凝土结构　　　　　　　cast-in-situ concrete structure
3. 装配式混凝土结构　　　　　　precast concrete structure
4. 缺陷　　　　　　　　　　　　defect
5. 严重缺陷　　　　　　　　　　serious defect
6. 一般缺陷　　　　　　　　　　common defect
7. 检验　　　　　　　　　　　　inspection
8. 检验批　　　　　　　　　　　inspection lot
9. 进场验收　　　　　　　　　　site acceptance
10. 结构性能检验　　　　　　　　inspection of structural performance
11. 结构实体检验　　　　　　　　entitative inspection of structure
12. 质量证明文件　　　　　　　　quality certificate document

六、Code for design of monolithic concrete residential buildings
《装配整体式混凝土居住建筑设计规程》DG/TJ 08—2071—2016

1. 预制混凝土构件　　　　　　　precast concrete component
2. 装配整体式混凝土结构　　　　monolithic precast concrete structure
3. 装配整体式混凝土框架结构　　monolithic precast concrete frame structure
4. 装配整体式混凝土异形柱框架结构　monolithic precast concrete frame structures with specially shaped columns
5. 装配整体式混凝土剪力墙结构　monolithic precast concrete shear wall structure
6. 装配整体式叠合剪力墙结构　　monolithic precast concrete composite shear wall structure
7. 预应力叠合空心楼板装配整体式剪力墙结构　monolithic precast concrete shear wall structure with prestressed composite slab
8. 装配整体式夹心保温剪力墙结构　monolithic precast concrete sandwich insulation shear wall structure
9. 预制外挂墙板　　　　　　　　precast concrete facade panel
10. 预制混凝土夹心保温外挂墙板　precast concrete sandwich facade panel
11. 连接件　　　　　　　　　　　connector
12. 钢筋套筒灌浆连接　　　　　　rebar splicing by grout-filled coupling sleeve
13. 金属波纹管浆锚搭接连接　　　rebar lapping in grout-filled hole formed with metal bellow

七、Code for acceptance of construction quality of steel structures
《钢结构工程施工质量验收规范》 GB 50205—2001

1. 零件 — part
2. 部件 — component
3. 构件 — element
4. 小拼单元 — the smallest assembled rigid unit
5. 中拼单元 — intermediate assembled structure
6. 高强度螺栓连接副 — set of high strength bolt
7. 抗滑移系数 — slip coefficient of faying surface
8. 预拼装 — test assembling
9. 空间刚度单元 — space rigid unit
10. 焊钉（栓钉）焊接 — stud welding
11. 环境温度 — ambient temperature

八、Code for design of steel structures
《钢结构设计标准》 GB 50017—2017

1. 强度 — strength
2. 承载能力 — load-carrying capacity
3. 脆断 — brittle fracture
4. 强度标准值 — characteristic value of strength
5. 强度设计值 — design value of strength
6. 一阶弹性分析 — first order elastic analysis
7. 二阶弹性分析 — second order elastic analysis
8. 屈曲 — buckling
9. 腹板屈曲后强度 — post-buckling strength of web plate
10. 通用高厚比 — normalized web slenderness
11. 整体稳定 — overall stability
12. 有效宽度 — effective width
13. 有效宽度系数 — effective width factor
14. 计算长度 — effective length
15. 长细比 — slenderness ratio
16. 换算长细比 — equivalent slenderness ratio
17. 支撑力 — nodal bracing force
18. 无支撑纯框架 — unbraced frame
19. 强支撑框架 — frame braced with strong bracing system
20. 弱支撑框架 — frame braced with weak bracing system
21. 摇摆柱 — leaning column
22. 柱腹板节点域 — panel zone of column web

23. 球形钢支座		spherical steel bearing
24. 橡胶支座		couposite rubber and steel support
25. 主管		chord member
26. 支管		bracing member
27. 间隙节点		gap joint
28. 搭接节点		overlap joint
29. 平面管节点		uniplanar joint
30. 空间管节点		multiplanar joint
31. 组合构件		built up member
32. 钢与混凝土组合梁		composite steel and concrete beam

九、Structural Welding Code-Steel
《钢结构焊接规范》GB 50661—2011

1. 消氢热处理　　　　　　　hydrogen relief heat treatment
2. 消应热处理　　　　　　　stress relief heat treatment
3. 过焊孔　　　　　　　　　weld access hole
4. 免予焊接工艺评定　　　　prequalification of WPS
5. 焊接环境温度　　　　　　temperature of welding circumstance
6. 药芯焊丝自保护焊　　　　flux cored wire selfshield welding
7. 检测　　　　　　　　　　testing
8. 检查　　　　　　　　　　inspection

十、Technical specification for high strength bolted connection of steel structure
《钢结构高强度螺栓连接技术规程》JGJ 82—2011

1. 高强度大六角头螺栓连接副　　heavy-hex high strength bolt assembly
2. 扭剪型高强度螺栓连接副关　　twist-off-type high strength bolt assembly
3. 摩擦面　　　　　　　　　　　faying surface
4. 预拉力（紧固轴力）　　　　　pre-tension
5. 摩擦型连接　　　　　　　　　friction-type joint
6. 承压型连接　　　　　　　　　bearing-type joint
7. 杠杆力（撬力）作用　　　　　prying action
8. 抗滑移系数　　　　　　　　　mean slip coefficient
9. 扭矩系数　　　　　　　　　　torque pretension coefficient
10. 栓焊并用连接　　　　　　　　connection of sharing on a shear load by bolts and welds
11. 栓焊混用连接　　　　　　　　joint with combined bolts and welds
12. 转角法　　　　　　　　　　　turn-of nut method

十一、Technical code of cold-formed thin-wall steel structures
《冷弯薄壁型钢结构技术规范》GB 50018—2002

1. 板件 — elements
2. 加劲板件 — stiffened elements
3. 部分加劲板件 — partially stiffened elements
4. 非加劲板件 — unstiffened elements
5. 均匀受压板件 — uniformly compressed elements
6. 非均匀受压板件 — non-uniformly compressed elements
7. 子板件 — sub-elements
8. 宽厚比 — width-to-thickness ratio
9. 有效宽厚比 — effective width-to-thickness ratio
10. 冷弯效应 — effect of cold forming
11. 受力蒙皮作用 — stressed skin action
12. 喇叭形焊缝 — flare groove welds

十二、Technical specification for space frame structures
《空间网格结构技术规程》JGJ 7—2010

1. 结构 — space frame; space latticed structure
2. 网架 — space truss; space grid
3. 交叉桁架体系 — intersecting lattice truss system
4. 四角锥体系 — square pyramid system
5. 三角锥体系 — triangular pyramid system
6. 组合网架 — composite space truss
7. 网壳 — latticed shell; reticulated shell
8. 球面网壳 — spherical latticed shell; braced dome
9. 圆柱面网壳 — cylindrical latticed shell; braced vault
10. 双曲抛物面网壳 — hyperbolic paraboloid latticed shell
11. 椭圆抛物面网壳 — elliptic paraboloid latticed shell
12. 联方网格 — lamella grid
13. 肋环型 — ribbed type
14. 肋环斜杆型 — ribbed type with diagonal bars (Schwedler dome)
15. 三向网格 — three-way grid
16. 扇形三向网格 — fan shape three-way grid (Kiewittdome)
17. 葵花形三向网格 — sunflower shape three-way grid
18. 短程线型 — geodesic type
19. 组合网壳 — composite latticed shell
20. 立体桁架 — spatial truss
21. 焊接空心球节点 — welded hollow spherical joint

Appendix, References & Acknowledgements | 193

22. 螺栓球节点　　　　　　　　　bolted spherical joint
23. 嵌入式毂节点　　　　　　　　embedded hub joint
24. 铸钢节点　　　　　　　　　　cast steel joint
25. 销轴节点　　　　　　　　　　pin axis joint

十三、*Code for design of steel-concrete composite bridges*
《钢-混凝土组合桥梁设计规范》GB 50917—2013

1. 钢混凝土组合梁　　　　　　　steel-concrete composite beam
2. 抗剪连接件　　　　　　　　　shear connector
3. 有效宽度　　　　　　　　　　effective width
4. 有效弹性模量　　　　　　　　effective modulus of elasticity
5. 材料强度标准值　　　　　　　characteristic value of material strength
6. 材料强度设计值　　　　　　　design value of material strength
7. 作用效应基本组合　　　　　　fundamental combination for action effects
8. 作用效应标准组合　　　　　　standard combination for action effects
9. 型钢混凝土组合结构　　　　　steel reinforced concrete composite structure

十四、*Code for design of the municipal bridge*
《城市桥梁设计规范》CJJ 11—2011

1. 可靠性　　　　　　　　　　　reliability
2. 可靠度　　　　　　　　　　　degree of reliability
3. 设计洪水频率　　　　　　　　design flood frequency
4. 设计基准期　　　　　　　　　design period
5. 设计使用年限　　　　　　　　design working life
6. 作用（荷载）　　　　　　　　action（load）
7. 永久作用　　　　　　　　　　permanent action
8. 可变作用　　　　　　　　　　variable action
9. 偶然作用　　　　　　　　　　accidental action
10. 作用效应　　　　　　　　　　effect of action
11. 作用效应的组合　　　　　　　combination for action effects
12. 设计状况　　　　　　　　　　design situation
13. 极限状态　　　　　　　　　　limit state
14. 承载能力极限状态　　　　　　ultimate limit states
15. 正常使用极限状态　　　　　　serviceability limit states
16. 安全等级　　　　　　　　　　safety classes
17. 高架桥　　　　　　　　　　　viaduct
18. 地下通道　　　　　　　　　　underpass
19. 小型车专用道路　　　　　　　compacted car-only road

十五、Code for seismic design of urban bridges
《城市桥梁抗震设计规范》CJJ 166—2011

1. 地震动参数区划	seismic ground motion parameter zoning
2. 抗震设防标准	seismic fortification criterion
3. 地震作用	earthquake action
4. E1 地震作用	earthquake action E1
5. E2 地震作用	earthquake action E2
6. 地震作用效应	seismic effect
7. 地震动参数	seismic ground motion parameter
8. 地震安全性评价	seismic safety assessment
9. 特征周期	characteristic period
10. 非一致地震动输入	nonuniform ground motion input
11. 场地土分类	site classification
12. 液化	liquefaction
13. 抗震概念设计	seismic conceptual design
14. 延性构件	ductile member
15. 能力保护设计方法	capacity protection design method
16. 能力保护构件	capacity protected member
17. 减隔震设计	seismic isolation design
18. $P\text{-}\Delta$ 效应	$P\text{-}\Delta$ effect

十六、Code for seismic design of buildings
《建筑抗震设计规范》GB 50011—2010

1. 抗震设防烈度	seismic precautionary intensity
2. 抗震设防标准	seismic precautionary criterion
3. 地震动参数区划图	seismic ground motion parameter zonation map
4. 地震作用	earthquake action
5. 设计地震动参数	design parameters of ground motion
6. 设计基本地震加速度	ground motion
7. 设计特征周期	design characteristic period of ground motion
8. 场地	site
9. 建筑抗震概念设计	seismic concept design of buildings
10. 抗震措施	seismic measures
11. 抗震构造措施	details of seismic design

十七、Standard for classification of seismic protection of building constructions
《建筑工程抗震设防分类标准》GB 50223—2008

1. 抗震设防分类	seismic fortification category for structures

2. 抗震设防烈度　　　　　　　　seismic fortification intensity
3. 抗震设防标准　　　　　　　　seismic fortification criterion

十八、Code for design of building foundation
《建筑地基基础设计规范》GB 50007—2011

1. 地基　　　　　　　　　　　　ground; foundation soils
2. 基础　　　　　　　　　　　　foundation
3. 地基承载力特征值　　　　　　characteristic value of subsoil bearing capacity
4. 重力密度（重度）　　　　　　gravity density; unit weight
5. 岩体结构面　　　　　　　　　rock discontinuity structural plane
6. 标准冻结深度　　　　　　　　standard frost penetration
7. 地基变形允许值　　　　　　　allowable subsoil deformation
8. 土岩组合地基　　　　　　　　soil rock composite ground
9. 地基处理　　　　　　　　　　ground treatment; ground improvement
10. 复合地基　　　　　　　　　 composite ground; composite foundation
11. 扩展基础　　　　　　　　　 spread foundation
12. 无筋扩展基础　　　　　　　 non-reinforced spread foundation
13. 桩基础　　　　　　　　　　 pile foundation
14. 支挡结构　　　　　　　　　 retaining structure
15. 基坑工程　　　　　　　　　 excavation engineering

十九、Technical code for building pile foundations
《建筑桩基技术规范》JGJ 94—2008

1. 桩基　　　　　　　　　　　　pile foundation
2. 复合桩基　　　　　　　　　　composite pile foundation
3. 基桩　　　　　　　　　　　　foundation pile
4. 复合基桩　　　　　　　　　　composite
5. 减沉复合疏桩基础　　　　　　composite foundation with settlement reducing piles
6. 单桩竖向极限承载力　　　　　ultimate vertical bearing capacity of a single pile
7. 极限侧阻力　　　　　　　　　ultimate shaft resistance
8. 极限端阻力　　　　　　　　　ultimate tip resistance
9. 单桩竖向承载力特征值　　　　characteristic value of the vertical bearing capacity of a single pile
10. 变刚度调平设计　　　　　　 optimized design of pile foundation stiffness to reduce differential settlement
11. 承台效应系数　　　　　　　 pile cap effect coefficient
12. 负摩阻力　　　　　　　　　 negative skin friction; negative shaft resistance
13. 下拉荷载　　　　　　　　　 downdrag
14. 土塞效应　　　　　　　　　 plugging effect

15. 灌注桩后注浆	post grouting for cast-in-situ pile
16. 桩基等效沉降系数	equivalent settlement coefficient for calculating settlement of pile foundations

二十、Technical code for ground treatment of buildings
《建筑地基处理技术规范》JGJ 79—2012

1. 地基处理	ground treatment; ground improvement
2. 复合地基	composite ground; composite foundation
3. 地站承载力特征值	characteristic value of subsoil bearing capacity
4. 换填垫层	replacement layer of compacted fill
5. 加筋垫层	replacement layer of tensile reinforcement
6. 预压地基	preloaded ground; preloaded foundation
7. 堆载预压	preloading with surcharge of fill
8. 真空预压	vacuum preloading
9. 压实地基	compacted ground; compacted fill
10. 夯实地基	rammed ground; rammed earth
11. 砂石桩复合地基	composite foundation with sandgravel columns
12. 水泥粉煤灰碎石桩复合地基	composite foundation with cement-fly ash gravel piles
13. 夯实水泥土桩复合地基	composite foundation with rammed soil-cement columns
14. 水泥土搅拌桩复合地基	composite foundation with cement deep mixed columns
15. 旋喷桩复合地基	composite foundation with jet grouting
16. 灰土桩复合地基	composite foundation with ted soil lime columns

二十一、Code for construction of building foundation engineering
《建筑地基基础工程施工规范》GB 51004—2015

1. 地基	subsoil
2. 基础	foundation
3. 复合地基	composite foundation
4. 桩基础	pile foundation
5. 强夯法	dynamic consolidation
6. 强夯置换法	dynamic replacement
7. 注浆法	grouting
8. 预压法	preloading
9. 振冲法	vibroflotation
10. 桩端后注浆灌注桩	post base grouting bored pile
11. 基坑工程	excavation engineering

12. 基坑支护结构		retaining structure of foundation pit
13. 咬合桩		secant pile
14. 型钢水泥土搅拌墙		steel and soil-cement mixed wall
15. 地下连续墙		diaphragm wall
16. 铣接头		cutter joint
17. 接头管（箱）		joint pipe (box)
18. 水泥土重力式挡墙		soil-cement gravity retaining wall
19. 土钉墙		soil-nailed wall
20. 逆作法		top-down method
21. 沉井		open caisson
22. 气压沉箱		pneumatic caisson
23. 地下水控制		groundwater control
24. 截水帷幕		curtain for cutting off water
25. 无筋扩展基础		non-reinforced spread foundation
26. 钢筋混凝土扩展基础		reinforced concrete spread foundation
27. 筏形与箱形基础		raft and box foundation
28. 盆式开挖		bermed excavation
29. 岛式开挖		island excavation
30. 锚杆（索）		anchor arm (rope)
31. 复合土钉墙支护		composite soil nailing wall

二十二、*Standard for acceptance of construction quality of building foundation*
《建筑地基基础工程施工质量验收标准》GB 50202—2018

1. 土工合成材料地基		geosynthetics foundation
2. 重锤夯实地基		heavy tamping foundation
3. 强夯地基		dynamic consolidation foundation
4. 注浆地基		grouting foundation
5. 预压地基		preloading foundation
6. 高压喷射注浆地基		jet grouting foundation
7. 水泥土搅拌桩地基		soil-cement mixed pile foundation
8. 土与灰土挤密桩地基		soil-lime compacted column
9. 水泥粉煤灰、碎石桩		cement flyash gravel pile
10. 锚杆静压桩		pressed pile by anchor rod

二十三、*Technical code for improvement of soil and foundation of existing buildings*
《既有建筑地基基础加固技术规范》JGJ 123—2012

1. 既有建筑		existing building
2. 地基基础加固		soil and foundation improvement

3. 既有建筑地基承载力特征值   characteristic value of sub-soil bearing capacity of existing buildings
4. 既有建筑单桩竖向承载力特征值   characteristic value of a single pile bearing capacity of existing buildings
5. 增层改造   vertical extension
6. 纠倾加固   improvement for tilt rectifying
7. 移位加固   improvement for building shifting
8. 托换加固   improvement for underpinning

二十四、*Technical specification for fiber reinforced concrete structures* 《纤维混凝土结构技术规程》CECS 38: 2004

1. 纤维混凝土   fiber reinforced concrete
2. 钢纤维   steel fiber
3. 异形钢纤维   special shaped steel fiber
4. 合成纤维   synthetic fiber
5. 等效直径   equivalent diameter
6. 纤维长径比   aspect ratio of fiber
7. 纤维体积率   fraction of fiber by volume
8. 韧性   toughness
9. 纤维混凝土结构   fiber reinforced concrete structures
10. 无筋纤维混凝土结构   fiber reinforced concrete structures without steel reinforcement
11. 钢筋纤维混凝土结构   fiber reinforced concrete structures with steel reinforcement
12. 预应力纤维混凝土结构   prestressed fiber reinforced concrete structures
13. 预应力钢纤维混凝土轨枕   prestressed concrete sleeper reinforced with steel fibers
14. 钢纤维局部增强预桩   precast concrete pile locally reinforced with steel fibers
15. 钢纤维混凝土叠合式受弯构件   composite flexural member of steel fiber reinforced concrete
16. 钢纤维部分增强钢筋混凝土构件   element partially strengthened with steel fibers
17. 层布式钢纤维混凝土复合路面   cement concrete with steel fiber reinforced concrete layers in the top and bottom
18. 钢纤维喷射混凝土   steel fiber reinforced shotcrete

二十五、*Standard test methods for fiber reinforced concrete* 《纤维混凝土试验方法标准》CECS 13: 2009

1. 钢纤维   steel fiber

2.	异形钢纤维	special shaped steel fiber
3.	合成纤维	synthetic fiber
4.	纤维混凝土	fiber reinforced concrete
5.	混杂纤维混凝土	hybrid fiber
6.	等效直径	equivalent diameter
7.	纤维长径比	aspect ratio of fiber
8.	纤维含量	content of fiber
9.	纤维体积率	fraction of fiber by volume
10.	韧性	toughness
11.	增强	strengthening
12.	增韧	toughening
13.	阻裂	crack resistance
14.	早龄期抗裂	crack resistance at early age

二十六、*Technical specification for strengthening concrete structures with Carbon fiber reinforced polymer laminate*
《碳纤维片材加固混凝土结构技术规程》CECS 146：2003

1.	碳纤维片材	carbon fiber reinforced polymer laminate
2.	碳纤维布	carbon fiber sheet
3.	碳纤维板	carbon fiber plate
4.	底层树脂	primer
5.	找平材料	putty fillers
6.	浸渍树脂	saturating resin
7.	粘结树脂	adhesives

二十七、*Standard for design of timber structures*
《木结构设计标准》GB 50005—2017

1.	木结构	timber structure
2.	原木	log
3.	锯材	sawn lumber
4.	方木	square timber
5.	板材	plank
6.	规格材	dimension lumber
7.	胶合材	glued lumber
8.	木材含水率	moisture content of wood
9.	顺纹	parallel to grain
10.	横纹	perpendicular to grain
11.	斜纹	at an angle to grain
12.	层板胶合木	glued laminated timber（Glulam）

13.	普通木结构	sawn and round timber structure
14.	轻型木结构	light wood frame construction
15.	墙骨柱	stud
16.	木材目测分级	visually stress-graded lumber
17.	木材机械分级	machine stress-rated lumber
18.	齿板	turns plate
19.	木基结构板材	wood-based structural-use panels
20.	轻型木结构剪力墙	shear wall of light wood frame construction

二十八、*Code for acceptance of constructional quality of masonry structures* 《砌体结构工程施工质量验收规范》 GB 50203—2011

1.	砌体结构	masonry structure
2.	配筋砌体	reinforced masonry
3.	块体	masonry units
4.	小型砌块	small block
5.	产品龄期	products age
6.	蒸压加气混凝土砌块专用砂浆	special mortar for autoclaved aerated concrete block
7.	预拌砂浆	ready-mixed mortar
8.	施工质量控制等级	category of construction quality control
9.	瞎缝	blind seam
10.	假缝	suppositious seam
11.	通缝	continuous seam
12.	相对含水率	comparatively percentage of moisture
13.	薄层砂浆砌筑法	the method of thin-layer mortar masonry
14.	芯柱	core column
15.	实体检测	in-situ inspection

二十九、*Code for design of masonry structures* 《砌体结构设计规范》 GB 50003—2011

1.	砌体结构	masonry structure
2.	配筋砌体结构	reinforced masonry structure
3.	配筋砌块砌体剪力墙结构	reinforced concrete masonry shear wall structure
4.	烧结普通砖	fired common brick
5.	烧结多孔砖	fired perforated brick
6.	蒸压灰砂普通砖	autoclaved sand-lime brick
7.	蒸压粉煤灰普通砖	autoclaved flyash-lime brick
8.	混凝土小型空心砌块	concrete small hollow block
9.	混凝土砖	concrete brick
10.	混凝土砌块(砖)专用砌筑砂浆	mortar for concrete small hollow block

11.	混凝土砌块灌孔混凝土	grout for concrete small hollow block
12.	蒸压灰砂普通砖、蒸压粉煤灰普通砖专用砌筑砂浆	mortar for autoclaved silicate brick
13.	带壁柱墙	pilastered wall
14.	混凝土构造柱	structural concrete column
15.	圈梁	ring beam
16.	墙梁	wall beam
17.	挑梁	cantilever beam
18.	设计使用年限	design working life
19.	房屋静力计算方案	static analysis scheme of building
20.	刚性方案	rigid analysis scheme
21.	刚弹性方案	rigid-elastic analysis scheme
22.	弹性方案	elastic analysis scheme
23.	上柔下刚多层房屋	upper flexible and lower rigid complex multistorey building
24.	屋盖、楼盖类别	types of roof or floor structure
25.	砌体墙、柱高厚比	ratio of height to sectional thickness of wall or column
26.	梁端有效支承长度	effective support length of beam end
27.	计算倾覆点	calculating overturning point
28.	伸缩缝	expansion and contraction joint
29.	控制缝	control joint
30.	施工质量控制等级	category of construction quality control
31.	约束砌体构件	confined masonry member
32.	框架填充墙	infilled wall in concrete frame structure
33.	夹心墙	cavity wall with insulation
34.	可调节拉结件	adjustable tie

三十、*Technical code for fire protection water supply and hydrant systems* 《消防给水及消火栓系统技术规范》 GB 50974—2014

1.	消防水源	fire water
2.	高压消防给水系统	constant high pressure fire protection water supply system
3.	临时高压消防给水系统	temporary high pressure protection water supply system
4.	低压消防给水系统	low pressure fire protection water supply system
5.	消防水池	fire reservoir
6.	高位消防水池	gravity fire reservoir
7.	高位消防水箱	elevated/gravity fire tank

8. 消火栓系统	hydrant systems/standpipe and hose systems
9. 湿式消火栓系统	wet hydrant system/ wet standpipe system
10. 干式消火栓系统	dry hydrant system/ dry standpipe system
11. 静水压力	static pressure
12. 动水压力	residual/ running pressure

三十一、*Code for design of automatic fore alarm system*
《火灾自动报警系统设计规范》 GB 50116—2013

1. 火灾自动报警系统	automatic fire alarm system
2. 报警区域	alarm zone
3. 探测区域	detection zone
4. 保护面积	monitoring area
5. 安装间距	installation spacing
6. 保护半径	monitoring radius
7. 联动控制信号	control signal to start and stop an automatic equipment
8. 联动反馈信号	feedback signal from automatic equipment
9. 联动触发信号	signal for logical program

三十二、*Code for fire protection design of buildings*
《建筑设计防火规范》 GB 50016—2014

1. 高层建筑	high-rise building
2. 裙房	podium
3. 重要公共建筑	important public building
4. 商业服务网点	commercial facilities
5. 高架仓库	high rack storage
6. 半地下室	semi-basement
7. 地下室	basement
8. 明火地点	open flame location
9. 散发火花地点	sparking site
10. 耐火极限	fire resistance rating
11. 防火隔墙	fire partition wall
12. 防火墙	fire wall
13. 避难层（间）	refuge floor（room）
14. 安全出口	safety exit
15. 封闭楼梯间	enclosed staircase
16. 防烟楼梯间	smoke-proof staircase
17. 避难走道	exit passageway
18. 闪点	flash point

19. 爆炸下限　　　　　　　　lower explosion limit
20. 沸溢性油品　　　　　　　boil- over oil
21. 防火间距　　　　　　　　fire separation distance
22. 防火分区　　　　　　　　fire compartment
23. 充实水柱　　　　　　　　full water spout

Appendix 3: Punctuation Marks and Typefaces
附录3：标点符号和字体

1. Punctuation marks 标点符号

  ( ) brackets, parentheses 圆括号
  < > angle brackets 尖括号
  [ ] square brackets 方括号
  { brace 大括号
  · bullet 着重号
  † dagger, obelisk 剑号
  ‡ double dagger 双剑号
  * asterisk/star 星号
  / solidus/oblique/slash/virgule 斜线
  § section 分节号
  ~ swung dash 代子号
  … ellipsis/suspension points 省略号
  ¶ paragraph 段落号

2. Typefaces 字体

	Serif 衬线体	Sans serif 无衬线体	slab Serif 粗衬线体
roman 罗马体	A	A	A
italic 斜体	*A*	*A*	*A*
bold face 黑体	**A**	**A**	**A**
light face 细体	A	A	A
swash capitals 花饰体大写字母			𝒜ℬ𝒞𝒟
upper case/capital letters 大写字母			A B C D
lower case/small letters 小写字母			a b c d e

Appendix 4: Weights and Measures
附录 4：度量

1. British and American, with metric equivalents 英美制与公制换算表

Linear measure	长度单位
1 inch（英寸）	=25.4 millimeters（毫米） exactly（精确值）
1 foot（英尺） =12 inches（英寸）	=0.3048 meter（米） exactly（精确值）
1 yard（码） =3 feet（英尺）	=0.9144 metre（米） exactly（精确值）
1 (statute) mile（法定英里） =1760 yards（码）	=1.609 kilometers（公里）
1 int. nautical mile （国际海里） =1.150779 miles（英里）	=1.852 kilometers（公里） exactly（精确值）

Square measure	面积单位
1 square inch（平方英寸）	=6.45 sq. centimeters（平方厘米）
1 square foot（平方英尺） =144 sq. in.（平方英寸）	=9.29 sq. decimeters（平方分米）
1 square yard（平方码） =9 sq. ft（平方英尺）	=0.836 sq. metre（平方米）
1 acre（英亩） =4840 sq. yd（平方码）	=0.405 hectare（公顷）
1 square mile（平方英里） =640 acres（英亩）	=259 hectares（公顷）

Cubic measure	体积单位
1 cubic inch（立方英寸）	=16.4 cu. centimeters（立方厘米）
1 cubic foot（立方英尺） =1728 cu. in.（立方英寸）	=0.0283 cu. meter（立方米）
1 cubic yard（立方码） =27 cu. ft（立方英尺）	=0.765 cu. metre（立方米）

Capacity measure	容积单位
British 英制	Metric 公制
1 fluid oz（液盎司） =1.8047 cu. in.（立方英寸）	=0.0284 litre（升）
1 gill（吉耳） =5 fluidoz（液盎司）	=0.1421 litre（升）
1 pint（品脱）=20 fluid oz（液盎司） =34.68 cu. in.（立方英寸）	=0.568 litre（升）
1 quart（夸脱）=2 pints（品脱）	=1.136 litres（升）
1 gallon（加仑） =4 quarts（夸脱）	=4.546 litres（升）
1 peck（配克）=2 gallons（加仑）	=9.092 litres（升）
1 bushel（蒲式耳） =4 pecks（配克）	=36.4 litres（升）

2. Metric, with British equivalents 公制与英制换算表

Linear measure	长度单位
1 millimetre（毫米）	=0.039 inch（英寸）
1 centimetre（厘米） =10mm（毫米）	=0.394 inch（英寸）
1 decimetre（分米） =10cm（厘米）	=3.94 inches（英寸）
1 metre（米） =100 cm（厘米）	=1.094 yards（码）
1 kilometre（公里） =1000m（米）	=0.6214 mile（英里）

Square measure	面积单位
1 square centimetre （平方厘米）	=0.155 sq. inch nq （平方英寸）
1 square metre（平方米） =10000 sq. cm（平方厘米）	=1.196 sq. yards
1 arce（公亩） =100 square metres （平方米）	=119.6 sq. yards （平方码）
1 hectare（公顷） =100 arces（公亩）	=2.471 acres（英亩）
1 square kilometre （平方英里） =100 hectares（公顷）	=0.386 sq. mile（平方公里）

Cubic measure	体积单位
1 cubic centimetre（立方厘米）	= 0.061 cu. inch（立方英寸）
1 cubic metre（立方米） =1000000 cu. cm（立方厘米）	= 1.308 cu. yards（立方码）
Capacity measure	容积单位
1 millilitre（毫升）	=0.002 pint（British）品脱（英制）
1 centilitre（厘升） =10ml（毫升）	=0.018 pint（品脱）
1 decilitre（分升） =10cl（厘升）	=0.176 pint（品脱）
1 litre（升） =1000ml（毫升）	=1.76 pints（品脱）
1 decalitre（十升） =10 l（升）	=2.20 gallons（加仑）
1 hectolitre（百升） =100l（升）	=2.75 bushels（蒲式耳）
1 kilolitre（千升） =1000l（升）	=3.44 quarters（夸脱）
Weight	重量
1 milligram（毫克）	=0.015 grain（格令）
1 centigram（厘克） =10mg（毫克）	=0.154 grain（格令）
1 decigram（分克） =100 mg（毫克）	=1.543 grains（格令）
1 gram（克） = 1000 mg（毫克）	=15.43 grains（格令）
1 decagram（十克） =10g（克）	=5.64 drams（打兰）
1 hectogram（百克） =100g（克）	=3.527 ounces（盎司）
1 kilogram（千克） =1000g（克）	=2.205 pounds（磅）
1 tonne（metric ton）吨（公吨） =1000kg（千克）	=0.984（long）ton（长）吨

Appendix 5: Numerals and Mathematical Symbols
附录5：数和数学符号

1. Numerals 数字

1/2: one half; a half; one over two
1/3: a third; one third; one over three
2/3: two thirds; two over three
1/4: a quarter; one quarter; one fourth; one over four
3/4: three quarters; three fourths
$2\frac{1}{2}$: two and a half
0.3: zero point three; O point three
0.03: zero point zero three; O point O three
0.67: zero point six seven; O point six seven
2%: two percent
2‰: two per mille
2/3m: two thirds of a meter
3/4km: three quarters of a kilometer
7.8m/s: seven point eight meters per second
15℃: fifteen degree Centigrade
34℉: thirty four degrees Fahrenheit
10kN: ten kilo Newton
20mm: twenty millimeters

2. Mathematical Symbols 数学符号

+   plus or positive   加或正
−   minus or negative   减或负
×   multiplied by   乘以
÷   divided by   除以
±   plus or minus, positive or negative   加或减，正或负
=   equal to   等于
≡   identically equal to; identically equals   恒等于
≠   not equal to   不等于
≈   is approximately equal to; approximately equals   约等于
∼   of the order of or similar to   约，近似
>   greater than   大于

$<$	less than	小于
$\not>$	not greater than	不大于
$\not<$	not less than	不小于
$\geqslant$	greater than or equal to	大于等于
$\leqslant$	less than or equal to	小于等于
$\sqrt{\ }$	square root	平方根
∞	infinity	无限大
$\propto$	proportional to	与…成正比
$\sum$	sum of …	之和
Δ	difference	差
$\therefore$	therefore	所以
$\angle$	angle	角
$//$	parallel to	平行于
$\perp$	prependicular to	垂直于
$:$	is to	比
$\lvert x \rvert$	the absolute value of x	x 的绝对值
b'	b prime	b 的求导
b''	b second prime	b 的二次求导
b_1	b subscript one	b 下标 1
b^2	b superscript two	b 的平方
$\mathrm{d}x$	Dee x; differential x	x 的微分
$\dfrac{\mathrm{d}^n y}{\mathrm{d}x^n}$	the n th derivative of y with respect to x	关于 y 的 n 阶导数
$\int$	integral	积分
$\int_a^b x$	integral between limits a and b	从 a 到 b 的积分
$(a+b)$	bracket a plus b bracket closed	括号里 $a+b$
$a:b$	the ratio of a to b	a 与 b 之比
x^2	x squared	x 的平方
x^3	x cubed	x 的三次方
$\dfrac{\partial y}{\partial x}$	the partial derivative of y with respect to x	y 相对于 x 的偏导数
$\sqrt{x}$	the square root of x	x 的平方根
$\log_n x$	log x to the base n	以 n 为底，x 的对数

References
参 考 文 献

[1] National Geographic Channel US. Engineering Connections—Sydney Opera House. [EB/OL]. http://www.imdb.com/title/tt1289217/. Episodes. 2007. Web.

[2] Darlow Smithson Productions UK. Seconds From Disaster [EB/OL]. http://natgeotv.com/uk/seconds-from-disaster/about/. 2004—2007. Web.

[3] PBS. NOVA. "Why the Towers fell." [EB/OL]. http://www.pbs.org/wgbh/nova/. 2010. Web.

[4] 沈祖炎. INTRODUCTION OF CIVIL ENGINEERING[M]. 北京：中国建筑工业出版社，2005.

[5] Nicholars Harris, Claire Aston and Emma Godfrey. Leap through Time - Earthquake[M]. Orpheus Books Ltd, 2 Church Green , Witney, OX28 4AW. 2004.

[6] 段兵延. 土木工程专业英语[M]. 武汉：武汉工业大学出版社，2002.

[7] 陈炳全. 建设与环境 CONSTRUCTION&ENVIRONMENT 建设及环境学院院刊 The Magazine of the Construction and Environment. Hong Kong Polytechnic University. 2011.

[8] 牛津英语联想词典[M]. 北京：商务印书馆，2003.

[9] 牛津实用英汉双解词典[M]. 北京：外语教学与研究出版社，2007.

[10] 英汉土木建筑大词典[M]. 北京：中国建筑工业出版社，1999.

[11] 牛津高阶英汉双解词典[M]. 北京：商务印书馆，2004.

[12] 刘雁滨，Wadeson. P. 实用口语突破[M]. 北京：外文出版社，2000.

[13] 刘润清，何福胜，张敬源. 当代研究生英语[M]. 北京：外语教学与研究出版社，2000.

[14] 鲁正. 土木工程专业英语[M]. 北京：机械工业出版社，2018.

[15] 宿晓萍. 土木工程专业英语[M]. 北京：北京大学出版社，2017.

[16] 俞家欢. 土木工程专业英语[M]. 北京：清华大学出版社，2017.

[17] 崔春义. 土木工程专业英语[M]. 北京：中国建筑工业出版社，2015.

[18] 苏小卒. 土木工程专业英语[M]. 上海：同济大学出版社，2015.

[19] 李锦辉，陈锐. 土木工程专业英语[M]. 上海：同济大学出版社，2012.

[20] 雷自学. 土木工程专业英语[M]. 北京：知识产权出版社，2010.

[21] 李亚东. 土木工程专业英语[M]. 成都：西南交通大学出版社，2005.

[22] 彭小悟. 基于堪萨斯城凯悦酒店坍塌事故的工程伦理学习[EB/OL]. https://www.jianshu.com/p/841b365ea4bb. 2017. Web.

[23] Leite, Fernanda, Akcamete, Asli, Akinci, Burcu, Atasoy, Guzide, Kiziltas, Semiha. Analysis of Modeling Effort and Impact of Different Levels of Detail in Building Information Models[J]. Automation in Construction, 2011.

[24] Smith, Deke. An Introduction to Building Information Modeling (BIM) [J]. Journal of Building Information Modeling, 2007.

[25] 袁佳珺. King's Cross Fire 全球首起重大地铁火灾：国王十字站大火[J] 上海安全生产，2010(12)：46-49.

[26] 袁佳珺. 掩藏15年的隐患——新加坡世界宾馆倒塌事故[J]. 上海安全生产，2010(3)：48-51.

Acknowledgements
致 谢

本书的出版是编者和研究生们共同努力的成果。除本书所列参考文献外，还广泛参考和借鉴了国内外大量的文献资料，在此对其相关作者一起表示衷心的感谢。此外，需要特别说明的是，研究生付敏完成第 4 章～第 9 章的视频英文记录和初步整理、郑挚完成第 2 章～第 12 章的中英文校对、徐乐完成第 2 章～第 12 章以及附录的专业词汇编选，华东理工大学夏延完成了第 2 章～第 9 章的图片处理，在此向他们辛勤付出表示衷心的感谢。

特别感谢武汉大学徐礼华教授、浙江大学徐世烺教授、湖南大学易伟建教授、湖北工业大学刘德富教授（校长）的勉励和支持。

特别感谢美国加州州立理工大学波莫纳分校（California State Polytechnic University, Pomona）贾旭东教授对本书给予非常宝贵的建议。

感谢中建三局马来西亚公司吴伟波先生对本书的关注和支持。感谢中国矿业大学李果副教授、武汉大学王若琳副教授、西南大学江胜华副教授、华中科技大学胡琴博士、武汉理工大学余泽川博士，以及同事李扬副教授、陈顺博士、陈娜博士等的鼓励和支持。

感谢教材分社副社长王跃编审（博士）给予的专业指导和真诚帮助，感谢赵莉编辑严谨的专业指导。感谢吴一红、杜国锋、朱爱珠等朋友给予的信任和支持。

编者
2019 年 1 月